AF735433

A FIELD GUIDE TO

Australian Frogs

A FIELD GUIDE TO

Australian Frogs

by

John Barker, Gordon Grigg
and Michael Tyler

Published by

Surrey Beatty & Sons

The National Library of Australia
Cataloguing-in-Publication entry:

A field guide to Australian frogs.

1995 ed.

Bibliography.

Includes Indexes.

ISBN 0 949324 61 2.

1. Frogs — Australia. — Identification. I. Barker, John, 1947–
II. Grigg, Gordon C. (Gordon Clifford), 1942–
III. Tyler, Michael J., 1937–

597.80994

Published October, 1995

Printed in Australia by

Surrey Beatty & Sons
43 Rickard Road, Chipping Norton, NSW 2170

Preface to the 1995 Edition

The eighteen years since the publication of the first edition of this Field Guide to Australian Frogs have seen three things which demand a revised edition. The first is that the first edition went out of print very rapidly but we were unable to persuade the then publishers, Rigby Limited, to reprint it. The second is the great increase in the level of interest in the natural history, ecology and conservation of the Australian fauna. The third and most important factor is that more than 50 new species of frogs have been described in that period, a 33 per cent increase in the number of known Australian species.

In the Preface to the 1977 edition J.B. and G.C.G. forecast that more species would be described, particularly in remote areas, and that taxonomic revisions would alter names. We foreshadowed the incorporation of such changes in a "future edition". This is it.

The aims of the present edition remain the same as the first: to enable easy identification of Australian frogs. The format remains that of a field guide and, indeed, is very similar to the original. We have tried to incorporate suggestions made by several colleagues and the biggest format change is in the distribution maps which, where appropriate, now provide information about the comparative distributions of closely related species. As before we have focussed on the use of easily seen characteristics for identification. There is a heavy reliance on colour photographs and we have tried to point out where two or more species may be confused and to offer diagnostic assistance. Photographs have usually been chosen with their usefulness for identification in mind, rather than photographic merit though we have, of course, tried to meet both aims.

The target audience remains the bright 11 year old, and up. We have tried to avoid technical terms and have provided a Glossary which defines those that cannot be avoided. Rather than include hundreds of citations to published work throughout the text, we include a bibliography to the sources of the information that we have incorporated.

The revision is being written at a time when there is concern that many species of frogs throughout the world are in decline. The insufficient baseline data, coupled with the serious technical difficulty of obtaining reliable data on the size and dynamics of frog populations, makes the issue a difficult one to address. Perhaps, with their thin, moist skins, frogs may be more easily affected by changes in the chemistry of their environment and could, therefore, be important indicator species, telling us about the health of our planet. Climatic changes are another distinct possibility. The identification of frogs, therefore, is of much more than just esoteric interest. We hope that this book, through providing an easy and portable guide to identification and natural history, will be useful to people who are already sensitive to the world of frogs and also draw others to the gentle art of frogging.

Acknowledgements

Gordon Grigg and John Barker wish to thank Michael Tyler for joining the team, an event which made the revision possible. Gordon wishes to record that the idea for a revision started with Barbara Smith, and that we are grateful to her (now). In some sections, particular assistance was rendered by experts and we wish to make special mention of Margaret Davies (*Uperoleia*), Michael Mahony and Dale Roberts (*Neobatrachus*) and Christina James (*Litoria bicolor* group and the section on frog taxonomy). Steve Richards and Keith McDonald provided many extremely constructive comments on the draft text. We are very grateful to all of the more than 20 individuals, identified throughout the book, who provided photographs. Photographs not credited otherwise are by Gordon Grigg. Maps were drawn by Kelly Fennell and the remaining artwork is by David Stanley (Figures 2–4, from the 1977 edition) or Lynn Pryor. We are grateful to the Departments of Zoology at The University of Adelaide and The University of Queensland, and the New South Wales National Parks and Wildlife Service, for support during the project. We also acknowledge the State and Territory authorities who provided permits enabling us to collect and study frogs. Gail Edwards keyed in some of the early drafts. The bulk of the word processing and much of the inter-city communication has been undertaken by Jan Mehaffey who has shown unbounded enthusiasm for the project throughout. Jan Grigg assisted considerably with proof reading and helping to assemble the final version. To all these people, plus those acknowledged in the 1977 edition, much of whose contributions live on in this one, and to any helpers whose names may have been missed inadvertantly, we extend a warm and hearty thanks.

John Barker, Sydney
Gordon Grigg, Brisbane
Michael Tyler, Adelaide
July 1994

Preface to the 1977 Edition

We have prepared this book primarily to enable identification of the 150 or so species of frogs found in Australia and to summarize some aspects of their biology. We have tried to minimize the use of technical terms so that the book may be useful to a wide audience. The book has been designed as a field guide and emphasis has therefore been given to descriptions of living specimens. Identification of a specimen is possible either from the series of colour photographs or from keys. Confirmation of identification and further information are provided by descriptions and notes on each species. We have tried to point out where confusion between similar species could occur and to summarize the diagnostic features in each case.

In recent years there have been a number of taxonomic changes and discoveries of new species. Wherever possible we have taken notice of these alterations and to the best of our knowledge the book is up-to-date. As more work is done, particularly in remote areas, further new species are bound to be described and there may well be further taxonomic changes. We hope to be able to incorporate any necessary changes in a future edition.

A number of people have been outstanding in their help and encouragement. Thanks are especially due to Dr H. G. Cogger, Dr A. K. Lee, and Mr Ken Simpson, whose constructive help with the text has been invaluable. Dr Grahame Webb provided many living specimens from Western Australia and specimens of *Pseudophryne corroboree* were generously donated by Mr Dane Wimbush. Mr John de Bavay kindly consented to our taking photographs of his specimens of *Kyarranus sphagnicolus*. We gratefully acknowledge permission to reproduce photographs by Dr H. G. Cogger, Professor A. R. Main, Mr O. E. S. Kelly, Mr Stan Breeden,

Dr M. J. Littlejohn, and Mr Roger Gaymer. All photographs not otherwise credited are by Gordon Grigg. Color-Master Pty Ltd of Willoughby, New South Wales, processed all film and we thank them for their cheerful and interested help. Dr Alex Ritchie made his specimens of fossil material available for photography. We would also like to thank Mrs W. Clayton for giving us specimens of *Neobatrachus pelobatoides*. We gratefully acknowledge the work of the artists who prepared material especially for this volume. Line diagrams are by David Stanley (Figures 4 and 8) and Diana Chrisafis (Figures 1, 3 and 6). Other diagrams are by Gordon Grigg. Considerable help was given by Miss Christine Lehmann in maintaining and cataloguing the large series of photographs and specimens. Numbers of friends, associates, and hangers-on have brightened many a field trip by their company. To all of these people we extend our warmest gratitude and hope that our efforts have justified their help and interest. Any errors and oversights which remain are our own. Funds for some of the work were provided by a research grant from the University of Sydney and this help is gratefully acknowledged.

Special thanks are due to Mrs Tessa Manalang whose patience and skill at the typewriter converted our untidy handwritten drafts to clean typescript.

If this book succeeds in fostering interest in our frog fauna, then the many thousands of kilometres we have travelled and the many hours of tramping about the countryside in the middle of wet nights will have been worthwhile.

J. Barker
G. Grigg

John Barker is a Technical Officer in the Scientific Research and Survey Branch of the New South Wales National Parks and Wildlife Service. Born in the USA of Scottish parents, he pursued the occupations of professional musician and industrial chemist before joining the School of Biological Sciences at The University of Sydney where he and Grigg teamed up to work on frogs. The author of numerous research papers, he has also worked on the effects of forest logging on fauna, diets of feral dogs and foxes, and conducted fauna surveys. His main current work is on koalas.

Gordon Grigg is Professor of Zoology at The University of Queensland. A Queenslander, he gained a PhD at the University of Oregon, USA and joined the staff at The University of Sydney in 1968. He was awarded a D.Sc. in 1987 for research in animal physiology. He took up his present appointment in 1989. He is the author or co-author of more than 100 research publications, dealing mainly with the physiology and/or ecology of vertebrate animals, especially crocodiles and kangaroos.

Michael Tyler, A.O., is Associate Professor of Zoology at the Univeristy of Adelaide and President of the Royal Zoological Society of South Australia. He has also had a long association with the South Australian Museum. Originally from the U.K., his first job in Australia was laying track on the Transcontinental Railway. The author or co-author of 14 other books and 250 articles, he specializes in the study of frogs. His current research interests range from fossils to the development of new drugs from skin secretions and the use of frogs as indicators of environmental pollution.

Contents

Chapter 1

Evolution, Taxonomy and Zoogeography of Frogs

The nature and origin of frogs

Frogs are a very successful but conservative group of animals. There are more than 4 000 species worldwide, occurring on every continent except Antarctica, and on many islands. Currently 204 species have been described from Australia, but at least 12 more have been discovered and await formal description and naming.

Frogs belong to a Class of vertebrate animals known as the Amphibia. This name derives from the Greek, *amphi* = both, *bios* = life, and refers to their dual life style which is usually aquatic during egg and larval stages and terrestrial during adulthood. Most species are dependent on an abundance of fresh water for breeding and this is, perhaps, the most significant factor controlling distribution. In those species adapted to life in arid areas, breeding occurs when rainstorms provide temporary ponds in which tadpoles can develop. In some wetter areas, particularly in rainforest or sphagnum and bog habitats, where the whole environment is perpetually moist, there are a few species which lay eggs away from water in a moist spot, and skip the free-living tadpole stage. Frogs are "cold-blooded" animals, having no physiological mechanisms for maintaining body temperature substantially different from that of the environment. Hence, frogs reach their greatest diversity in the tropics. Despite the restrictions of water and temperature, frogs have adapted to a great variety of ecological situations and have successfully colonized seemingly unsuitable habitats such as deserts and cold high mountains. The marine environment, however, has always been barred to them, presumably because of their permeable skin which affords no protection against water loss by osmosis to the more concentrated salty external medium. One exception is the celebrated Crab-eating Frog *Rana cancrivora* of South-East Asia, which is able to tolerate limited exposure to salt water in the intertidal zone. Cane Toads *Bufo marinus* (see p. 379) also have some tolerance to salty water, up to approximately 40 per cent sea water. Frogs exhibit a great diversity of breeding habits and behaviour, and of particular interest is the vocalization by the males to attract females of the same species. Correlated with the demands of their environments are diverse patterns of larval development including, in some instances, elaborate parental care of the young.

Despite considerable diversity in colouration, breeding habits, abilities to cope in different habitats and body size, all frogs are similar in overall body form. The skeleton, muscles and viscera (guts) of the tiny *Pseudophryne* species for example (p. 263) have the same general pattern as the world's largest species, the West African Goliath Frog *Conraua goliath*.

Frogs are descended from the first vertebrates to leave the water and walk on land. Among these first land vertebrates were the labyrinthodonts, some of which grew to about three metres in length. They were the first amphibians and arose in the Devonian period about 400 million years ago. Reptiles are descended from labyrinthodont-like creatures, and subsequently birds and mammals were derived from different groups of reptiles.

The modern Amphibia survive only as three separate groups (Orders): the tropical Gymnophiona (caecilians, Plate 1) which are burrowing, worm-like animals with a patchy geographic distribution across the world's tropical areas; the tailed Caudata (newts and salamanders, Plates 2 and 3), which occur widely in the northern hemisphere and in South America; and the tail-less Anura (frogs and toads), found on all continents except Antarctica.

Evolution of modern frogs

It appears that up to and including the Triassic period, the fossil record shows only tailed amphibians which foreshadow anuran structure. Fossils of frogs appear first in the Jurassic period of North and South America and Europe, 280 million years ago. Some but not all of the fossil genera are able to be assigned to the same families as living genera. It seems therefore that the primary evolution of many anuran families was completed by the Jurassic.

PLATE 1. Female caecilian *Grandisonia sechellensis* with eggs; Seychelles *(Photo courtesy Roger Gaymer).*

PLATE 2. *Ambystoma tigrinum* Tiger Salamander; eastern USA *(Specimen courtesy Taronga Zoo).*

PLATE 3. *Taricha torosa* Californian Newt *(Specimen courtesy Taronga Zoo).*

The evolutionary path of modern frogs involved a process of shortening of the body and loss of the tail, and a single cast of an amphibian found in Madagascar is the only fossil found so far which links modern frogs and their ancestors. Named *Triadobatrachus masenotti*, it has conventional anuran skull bones and limb structure, but has a longer body with more vertebrae, as well as vertebrae in the tail. Its classification has been debated vigorously, but the majority view is that it represents a transitional stock from which the modern frogs evolved. It is, perhaps, best viewed as a "pre-anuran".

Knowledge of the fossil record of frogs has developed considerably over the past 20 years but there remains a large number of gaps and it is too early to say just where the first frog evolved. However, it is likely to have been somewhere in the southern hemisphere.

There are particularly good specimens and casts in Europe (Plate 4) and North and South America, and excellent fossil tadpoles have been found in Israel. The Australian fossil record is confined to isolated bones found in caves, dry lake beds or embedded in fresh-water limestones. The first fossil frog found in Australia was a portion of a pelvis from Lake Palankarinna in South Australia reported in 1974. Subsequently, fossils

PLATE 4. Fossil frog, *Rana pueyoi;* from central Spain, Miocene. Carbon stains show the body outline *(Specimen courtesy Alex Ritchie).*

representing at least 2 000 individuals and more than 30 species have been located at many mainland sites including the famous Riversleigh site in north-west Queensland and at King Island in Bass Strait.

The oldest Australian frog fossils so far recognized are from Murgon in south-east Queensland, from a freshwater clay deposit dated at 54.5 million years. *Lechriodus casca* was reported from the site in 1993 and *Litoria* and *Limnodynastes* species await description.

Considerable progress has been made in gaining some understanding of the Australian fossil record since our first fossil frog was described 20 years ago. Nine genera are known, all representing either of the two dominant modern families Hylidae and Myobatrachidae (= Leptodactylidae, see later), and only one genus *Australobatrachus,* is extinct.

Amphibians and continental drift

After long debate, the theory of continental drift is now accepted. The theory is a very exciting one, proposing that the world's land was once a single unit from which the southern half (Gondwana) incorporating Africa, India, Australia, South America, and Antarctica split off and then fragmented into the separate continents which slowly "drifted" to their present positions some time since the early Cretaceous period (100 million years ago). Long before there was convincing geological and palaeomagnetic evidence in support of this theory, attention had been drawn to the way the southern continents in particular, could be fitted together like pieces of a jig-saw, and to the similarities between many of the rock types and plant and animal fossils on adjacent but distant continents. This similarity is sometimes repeated in their modern faunas. As more and more data have been collected, the idea has become far more compelling. For example, palaeontologists have collected fossil labyrinthodonts in Antarctica. Because it is thought that labyrinthodonts, like modern Amphibia, could not tolerate exposure to salt water, the Antarctic labyrinthodonts could not have arrived there by sea, so Antarctica must once have been joined to other land masses, and enjoyed a much more temperate climate. The relevance of ideas about continental drift to Australian frogs is that the modern amphibian faunas in Australia and South America are similar in many respects. No fossil frogs have yet been found in Antarctica but it seems safe to assume that frogs did occur there and that ancestral stocks crossed between Australia and South America. This so-called Gondwana connection must be borne in mind in studies of anuran zoogeography.

Classification of frogs

The fundamental unit in animal classification is the "species". Many attempts have been made to accurately define a species, but it is difficult to establish hard and fast rules. In general terms, the members of a species look alike, have similar habits and, most importantly, they interbreed to produce fertile young. Similar species are grouped into a "genus" (plural: genera). Each species has a two-part "scientific" name consisting of the generic name followed by the specific name, usually written in italics. Thus the Green Tree Frog is called *Litoria caerulea,* whereas Peron's Tree Frog, *Litoria peronii,* is another species in the same genus. Diagnosis of a species is sometimes difficult to achieve, but it is often much easier than the grouping of species into genera, genera into families, families into orders, and so on. In fact, groupings at this level of taxonomy tend to be fairly arbitrary and depend on the (frequently differing) opinions of competent taxonomists (as scientists who study animal classification are usually known). The taxonomic classification of *Litoria peronii* is as follows:

Phylum:	Chordata
Sub-Phylum:	Vertebrata
Class:	Amphibia
Order:	Anura
Family:	Hylidae
Genus:	*Litoria*
Species:	*peronii*

In some cases, taxonomists feel that further subdivision is required. Thus families may be divided into two or more sub-families and species may be divided into sub-species. For example, Australian frogs of the family Myobatrachidae (= Leptodactylidae of some authors, including M.J.T.) are grouped into two sub-families, the Limnodynastinae (which have a particular type of specialization of the throat musculature) and the Myobatrachinae (which lack the specialization). A widespread species within the Limnodynastinae is *Limnodynastes dumerilii* (p. 181) and on the basis of morphology (structure) and other subtle differences it is possible to recognize a number of sub-species. These sub-species are given three-part names, for example, *Limnodynastes dumerilii dumerilii*, *Limnodynastes dumerilii insularis*, and *Limnodynastes dumerilii grayi*.

Quite frequently, but only where confusion cannot possibly occur, generic and specific names are abbreviated: *L. peronii* = *Litoria peronii* (but not if confusion with *Limnodynastes peronii* could occur); *L. d. grayi* = *Limnodynastes dumerilii grayi*.

The name (or names) of the taxonomist(s) who originally described a species are usually given after the species name, for example, *Limnodynastes fletcheri* Boulenger. Where the species is now placed in a genus other than that in which it was originally described, convention demands that the author's name be placed in parentheses, for example *Litoria rothii* (de Vis), which until 1971 was known as *Hyla rothii* de Vis.

The tools of the frog taxonomist

Clear-cut features of morphology are often sufficient to define a species. There are, however, many examples among Australian frogs where groups of species within one genus cannot be confidently separated in this way. It often happens that there are enough morphological differences between frogs from different localities to suggest that they may be separate species, yet the differences are slight and perhaps variable. In such circumstances, the taxonomist must look beyond purely morphological characteristics to decide whether one or more species is present. As we shall see, attempts to answer such questions have turned many a frog taxonomist into an experimental zoologist, sound analyst, or biochemist.

If morphological features are insufficient to allow a decision about the status of the "species" in question, the taxonomist must turn to other criteria. The crux of the species definition lies in the idea of a species as a breeding group. That is, any member of the group can mate with any opposite-sexed member of the same group and produce normal and fertile offspring, but the group is reproductively isolated from other species. If it can be shown that two "species" in question cannot and do not interbreed, then the significance of the distinction suggested by minor morphological differences is confirmed. Sometimes species can be defined in this way when morphological differences are lacking. Such species are called "cryptic" or "sibling" species because the differences are hidden to our eyes. To related species, however, differences in behaviour or advertisement calls may provide a unique and conspicuous identity.

A common way in which the question of interbreeding has been tackled is by the use of artificial cross-fertilization experiments. Such experiments are possible in frogs because the eggs are fertilized and develop externally. Therefore it is a simple matter to set up an artificial cross by taking eggs from a gravid female, adding them to a suspension of sperm, and observing subsequent events. In many cases fertilization occurs but results in aberrant development. This gives clear evidence of genetic incompatibility. In some cases, however, development may appear normal and this may prove difficult to evaluate because hybrids between even distantly-related species can result from such artificial crosses. It is probable that the hybrid offspring from such an artificial cross would be infertile, but this is difficult to determine due to the problems encountered in rearing frogs to maturity.

Genetics is rapidly becoming an essential tool for taxonomy, systematics, population biology and conservation biology. Genetic techniques provide a rapid and precise way to recognize cryptic species, hybrids and hybrid zones, as well as to quantify population variation divergence, and the effects of population subdivision and habitat fragmentation.

The genetic make-up of an organism can be studied at several levels of resolution. At a gross level there is "karyology". Karyology is the study of the number (denoted as "n" from sperm and ova, and "2n" from body cells), size, and morphology of the chromosomes of each cell. Chromosomes are visible under a light microscope in cells at a certain stage of cell division and can easily be prepared from a blood smear. Australian tree frogs and microhylids typically have 13 chromosome pairs (i.e., 2n = 26). The White-lipped Tree Frog *Litoria infrafrenata* is an exception with only 2n = 24 chromosomes. Ground-dwelling myobatrachids are characteristically 2n = 24, with a few exceptions: *Limnodynastes salmini*, the *L. dumerilii* group, *L. ornatus*, *L. dorsalis*, *L. terraereginae* and *L. interioris* having 2n = 22. Of the 10 recognized species of *Neobatrachus*, which normally have 2n = 24, *N. aquilonius*, *N. centralis*, *N. kunapalari*, *N. sudelli* and *N. sutor*, are unusual in that they are tetraploid, i.e., they have twice the normal complement of chromosomes, with 2n = 48.

More detailed genetic approaches come from the field of molecular genetics. In molecular genetics, the genetic material that chromosomes are made of (the "nucleic acids", or "DNA") is the focus of study. Particular sites along the DNA chains are called "genes" which are the blueprints for proteins. That is, genes instruct the cell how to assemble the various proteins essential for the cells and, in turn, the whole body's metabolism, development and differentiation. Genes are modified in a unique way during speciation and the divergence of populations. In fact, any evolutionary divergence or adaptation is ultimately genetically mediated. Different species and populations will therefore show varying degrees of genetic differences with some genes showing little variation, while others show a great deal. It is these differences that molecular genetic studies can detect.

Molecular genetics can therefore focus either on the various proteins, as a reflection of the genes, or on the genes themselves. A widely used technique for studying proteins is "allozyme electrophoresis". The essence of this method is that different proteins, which separate at different rates due to their mobility when placed in an electric field, can be used as gene markers. Systematic, taxonomic, ecological and conservation studies of many species of Australian frogs are coming to rely heavily on allozyme electrophoresis.

Following a similar principle, when strands of DNA have been cut with site specific enzymes, the resulting pieces or "fragments", can be separated via an electric current. Small fragments move faster through a matrix than do larger fragments. Banding patterns produced from allozyme electrophoresis and DNA analysis are then compared. Distinctive banding patterns are usually diagnostic of species status or significant geographical or reproductive isolation.

The most detailed molecular genetic approach is to determine the actual sequence of the nucleic acids making up specific genes of the DNA of an individual animal. The type and number of differences between DNA sequences gives precise information regarding the evolutionary history of the group being studied. A current systematic revision of the genus *Litoria* relies heavily on such molecular genetic techniques. Herpetology has greatly benefited from collaboration between molecular genetic methods and other fields of science. For example, *Litoria xanthomera*, the Orange-thighed Tree Frog, was recognized as a distinct species from its southern relative *Litoria chloris*, the Red-eyed Tree Frog, on the basis of morphology and allozyme electrophoresis.

Another approach is the use of techniques for analysing the advertisement calls of adult males. Typically, most frogs have a distinct call to which only the females of the same species will respond. Thus, two closely related species, breeding in the same pond at the same time of year, will not interbreed because the females respond only to the males of their own species. Mechanisms such as this are sometimes called pre-mating isolating mechanisms and can be seen as part of a specific mate recognition system which effectively prevents mating between different species. Frog advertisement calls are perhaps the best known example. Mis-matings, which occur occasionally, either fail to produce eggs or else produce inviable or infertile hybrid offspring. This is wasteful of the population's reproductive effort so it is easy to imagine natural selection favouring the survival of individuals whose genetic make-up supports behaviours which minimize the chance of mis-matings. The site from which males call, whether in the water, on the bank, or in low vegetation, and the time of year when breeding occurs, are commonly species-specific and may reinforce specific mate recognition cues. Because frog calls are easy to recognize, record, and analyse, much attention has been focused on male calls as a taxonomic tool.

Characteristics of the call which are easily analysed are the dominant frequencies (those with most sound energy), call duration, amplitude modulation, and pulse repetition rate. These may be determined by analysis of recorded calls on an oscilloscope or on a sonagraph which draws a graph (sonogram) of the sound frequency spectrum against time (Fig. 1). In the most modern approach, the sound may be digitized by computer and analysed quantitatively in the most minute detail. Apart from description of the sonic characteristics of the mating calls, it has often been possible to test the response of the female to a particular call, and even to test her discrimination between calls from closely related species played simultaneously. These experiments are usually carried out in a sound chamber or arena with a loudspeaker at each end. A female in breeding condition is placed in the middle. Her approach to the speaker transmitting the call of the "correct" species shows the role of the call as an attractant and, further, emphasizes the role of the mating call in mate recognition.

There are many examples among Australian frogs where call analysis and other techniques have not yet been applied and it is certain that future research will lead to the recognition of additional cryptic species, as well as better understanding of the evolutionary relationships within taxonomic groupings.

To a large extent, the process of speciation is the process by which reproductive isolation from closely related species evolves. The exact processes by which new species evolve are not fully understood, but genetic divergence as a consequence of adaptation to environmental changes during a period of geographic isolation is likely to be a common way for speciation to occur. Recognizable geographic races which are not reproductively isolated are likely to be in the process of evolving into separate species and taxonomists sometimes denote well-defined geographic races as sub-species (e.g., *Litoria verreauxii verreauxii* and *L. v. alpina*, see pp. 91, 92).

Zoogeography of Australian frogs

Only about 200 of the world's 4 000 known frog species are found in Australia. This relatively impoverished fauna may be related to the aridity of most of the continent. Lack of species numbers is, however, compensated for by the diversity of habits seen among Australian frogs. Large variations in rainfall and temperature, depending on latitude and elevation, have resulted in considerable habitat diversity.

The Australian continent has few obvious natural barriers such as high mountain ranges or extensive inland seas. Accordingly, some of the more adaptable species have been able to disperse over very wide areas. Bass Strait is a significant barrier, although the frog fauna in Tasmania is similar to that in Victoria but with fewer species. This is in accord with Tasmania's land connection with the mainland until about 40 000 years ago. Tracts of arid country often act as barriers to frog dispersal. For example, the coastal desert adjacent to the Great Australian Bight now isolates the damp coastal forests in the south-east and south-west. At the species level, the frog faunas of these two areas are quite different, a heritage of long isolation. On the other hand, common ancestry is indicated by the close

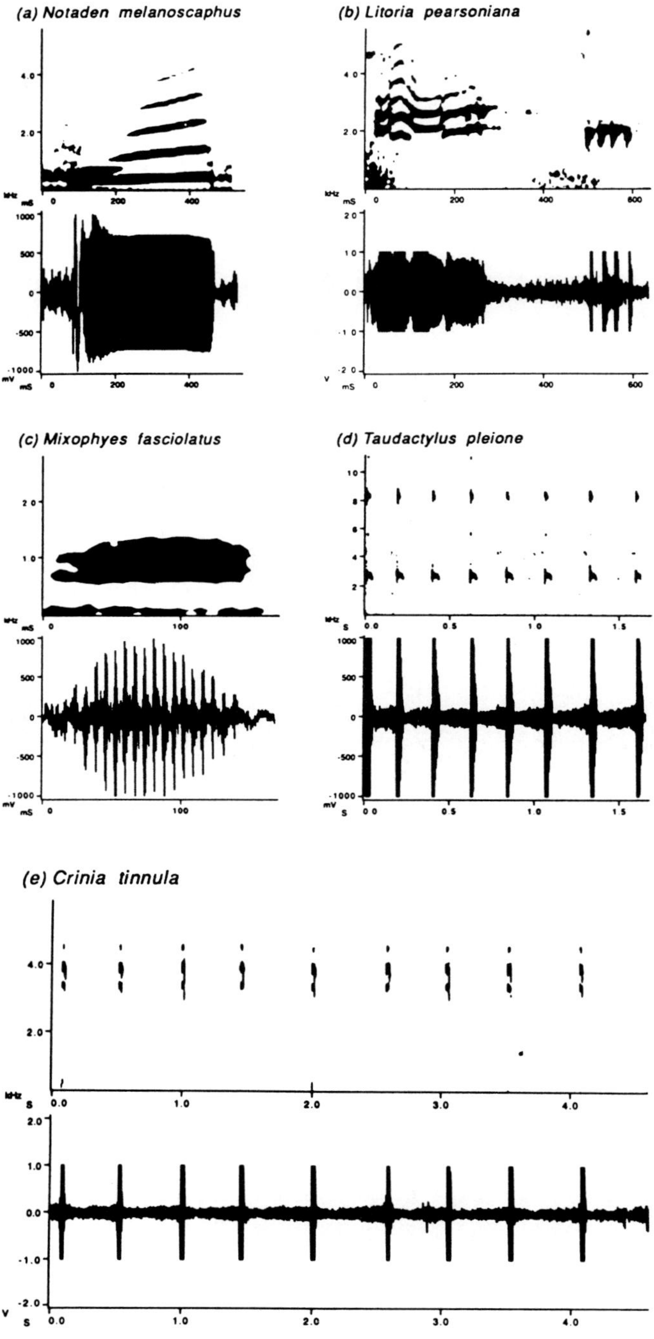
(a) Notaden melanoscaphus
(b) Litoria pearsoniana
(c) Mixophyes fasciolatus
(d) Taudactylus pleione
(e) Crinia tinnula
kHz
mS
mV
V
S

(f) Pseudophryne coriacea

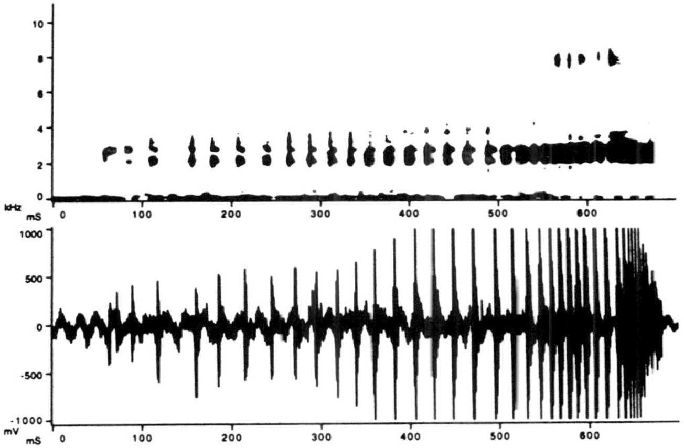

(g) Adelotus brevis

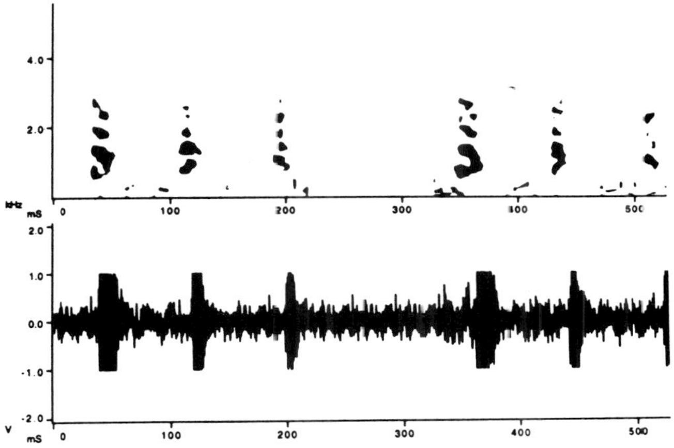

Fig. 1 (a–g). A male frog's advertisement call is usually diagnostic for the species, but is often very difficult to describe in words. Sonograms and oscillograms depict calls graphically and allow detailed description of a call's temporal characteristics, sound frequency composition and pattern of sound intensity which may allow discrimination between closely related species. Examples of sonograms (upper panels) and oscillograms (lower panels) from seven species are shown here. Horizontal axes show time in either seconds or milliseconds, vertical axes show frequency (pitch) in KHz (sonograms) or signal amplitude in volts/millivolts (oscillograms). Note that the maximum amplitude of oscillograms is truncated at ±1 volt in these examples. Phonetic descriptions of each call are given with each species description.

similarity between these two faunas at generic level. This affinity reflects the existence of a wet belt connecting south-east and south-west in times long past. The latitudinal and altitudinal span of the Great Dividing Range has led to great diversity of habitats and this diversity is echoed in the number of species which occur in the vicinity.

Australia can be divided into four zoogeographic provinces: Torresian, Bassian, Eyrean, and South Western. Sir Baldwin Spencer proposed these divisions late in the 19th century and they were revised in 1957 by Serventy and Whittell, working on birds. The anuran fauna fits this scheme relatively well except that the boundary between Bassian and Torresian provinces would be better situated in southern Queensland instead of at the Clarence River.

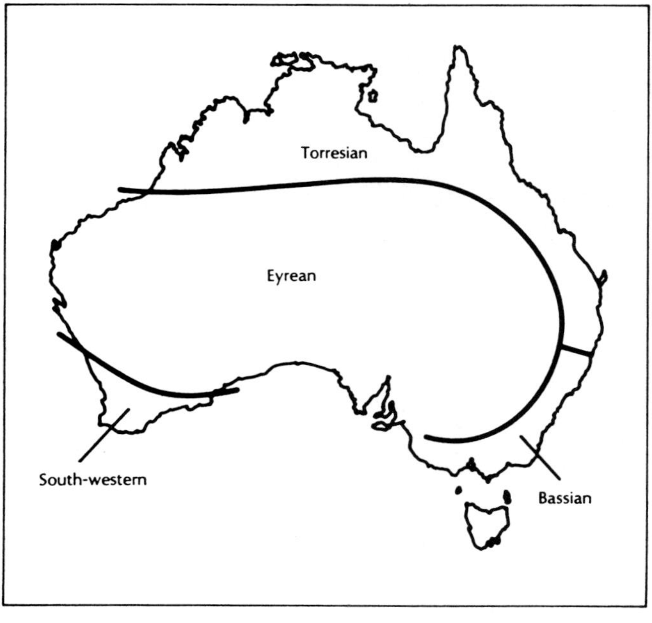

Fig. 2. Zoogeographic sub-regions.

Chapter 2

Biology of Frogs

Ode to Frog

What a wonderful bird the frog are.
When he sit, he stand; almost.
When he hop, he fly; almost.
He ain't got no sense, hardly,
He ain't got no tail, hardly, either.
When he sit, he sit on
What he ain't got — almost.

Anon.

Frogs characteristically have broad bodies and broad heads, no distinct neck, no tail in the adult stage, and hind limbs which are always larger than the front legs. The mouth is always at the very front of the body and the cloaca (common exterior opening for release of faeces, urine, and sperm or eggs) at the rear. In almost all species the eyes are prominent. Skeletal features are broadly similar throughout, although variations do occur and these are used for classification at higher levels of taxonomy.

Locomotion

Frogs are presumably descended from tailed Amphibia whose locomotion on land was accomplished by walking in a similar manner to the modern newts and salamanders. Undoubtedly much of the evolutionary success of frogs derives from their ability to jump. Their tail has been lost and the vertebral column shortened and strengthened; the hind limbs are elongated and strong, and provide motive force for the leap, while the front limbs and shoulder girdle absorb the stresses imposed at each landing. The ability of many frogs to leap a metre or more has obvious advantages for the capture of prey and the rapid escape from would be predators. The development of adhesive pads on fingers and toes has given many frogs access to insect life among the trees and rocks. However, not all frogs hop. Some of the smaller ground frogs, such as *Pseudophryne* and *Uperoleia* species, have only short legs and normally adopt a walking gait although they can make short hops, whilst the globular *Notaden* species can run.

A significant percentage of Australian frogs burrow to avoid harsh environmental conditions. The hind legs, often bearing hardened digging tubercles (swellings) on the feet, scoop soil from below the frog as it retreats backwards into soft soil. Exceptions are the Western Australian species *Myobatrachus gouldii* (p. 356) and *Arenophryne rotunda* (p. 354) which burrow head first.

Most frogs are equally at home in the water, where locomotion is achieved by simultaneous movements

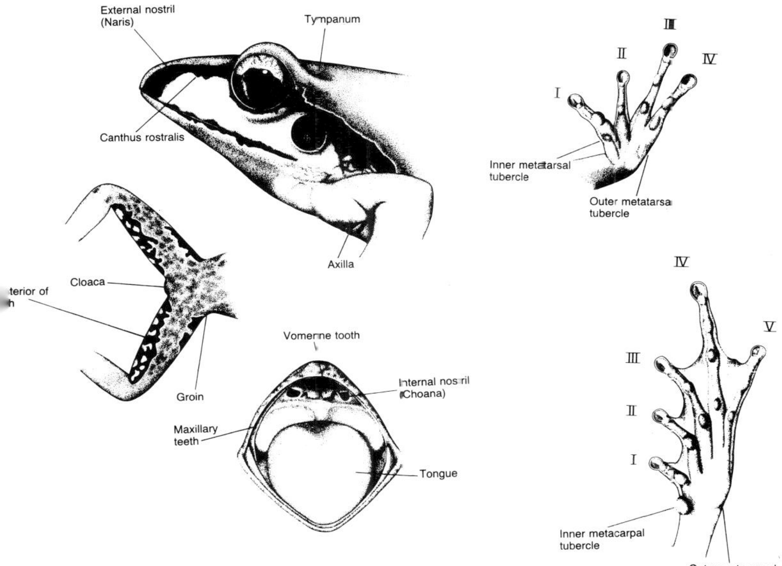

Fig. 3. Parts of a frog (Artwork by David Stanley).

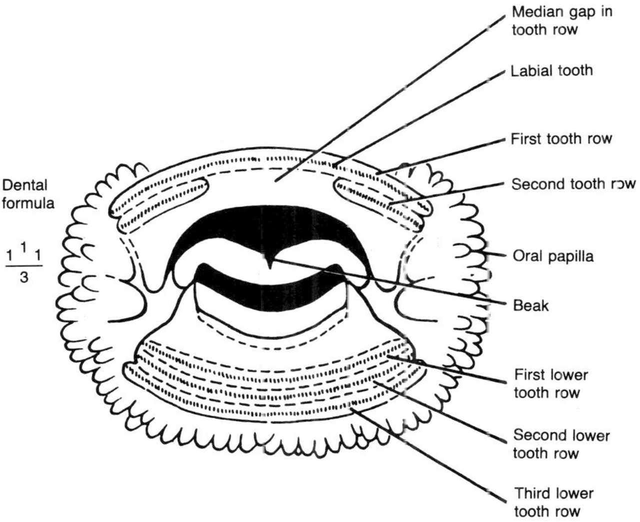

Fig. 4. Mouthparts of a generalized tadpole.

of the hind legs — the "frog-kick", or using "dog-paddle". Extensive webbing between the toes of most aquatic species enhances the effectiveness of this.

Skin

The skin is important as an organ of gas and water exchange, defence, camouflage, sensation, and as a buffer between the frog and its environment. The surface cells of the skin are compressed and keratinized (horny), and the skin is shed periodically and eaten in one piece. There are several kinds of glands in the skin, of which the most common are mucous glands secreting a clear watery fluid which moistens the skin and facilitates gas exchange, and granular glands which secrete poisons which serve as defence against predation or contain antibiotic compounds that combat skin infection. In some species, concentrations of these glands form large elevated areas. Those behind the eye are called parotoid glands. In toads of the genus *Bufo*, the parotoid glands secrete copious quantities of noxious neurotoxic material which can damage the eyes, produce numbness of the mouth and, if eaten by a predator, can cause rapid death.

Chromatophores (pigment cells), as well as layers of pigment between the cellular layers, give the frog its characteristic coloration. In some species, the pattern of coloration is cryptic so that the frog is well camouflaged (Plate 5), but others are brightly coloured and this is thought to be a warning of chemical secretions which make the frog distasteful to predators. Many species of frog are able to change colour as a result of the migration of pigment granules within the chromatophores. Colour changes are effected by hormones in response to various factors such as light intensity, temperature, humidity and stress.

The skin of frogs is permeable, allowing water movement in both directions. This means that as well as losing water quickly in dry conditions, frogs can regain water rapidly from a moist surface. Many Australian burrowing frogs are able to retard water loss by secreting a cocoon around themselves in the burrow. This cocoon is formed from the shedding of outer layers of the skin, and dormant frogs in their cocoons are able to survive for many months on food and water stored within the body. The permeability of the skin may make frogs more vulnerable to chemical pollution.

Senses

Sight, touch, hearing and olfactory (smell) senses are particularly well developed in frogs and toads. The sense of touch is achieved by numerous nerve endings in the skin, particularly on the head, tongue, and on the feet.

The importance of having good vision is obvious when one considers their modes of locomotion and prey capture. In most species the eyes are large and protruding, permitting a good capacity for binocular vision. In strong light the pupils contract markedly whereas at night pupil dilation allows good night vision. The lens of the eye is spherical, and focusing is achieved by muscles which move the lens forwards or backwards. Cones are present in the retina which suggests some degree of colour vision.

Good night vision is important, too, for mutual identification between conspecifics during courtship, much of which occurs at night. In some species, visual signalling by foot-waving has been described, at distances of one metre or so. Originally it was thought that this may be a substitute for vocalizations in frogs living in noisy environments, such as by rushing streams. However, Steve Richards and Christina James have observed foot-flagging in *Litoria eucnemis/genimaculata*, *L. nannotis*, *L. rheocola* and *L. fallax*, all of which use vocalization to attract conspecific females. They saw the displays only in the presence of rival males, and concluded that the behaviour relates more to male-male aggression than to courtship. It seems likely that foot-flagging may be an important supplement to vocalizations in many species and further studies are needed. It may be that the increasing availability of night-vision equipment will facilitate more work on this.

Hearing is well developed in anurans and is of great importance in reproductive behaviour, for sound is one of the stimuli which leads to aggregations of breeding frogs and effects mate recognition. The outer ear consists of the tympanic membrane or eardrum which is clearly visible in most *Litoria* species and some myobatrachids. In many myobatrachids, however, the membrane is covered by the skin of the head and is indistinct. The ears of frogs have a wide frequency response, from 50 to 10 000 hertz, but that of a particular species may be much narrower and there is often a "tuning" to increased sensitivity at the frequencies characteristic of the relevant mating call. In *Pseudophryne*, hearing apparatus is lacking but species in this genus do vocalize so there may be an alternative mechanism for conducting sound to the inner ear, perhaps via the bones of the skull.

Feeding and digestion

Feeding habits change dramatically during the life cycle. Tadpoles are omnivorous, grazing on algae or animal matter, but they can swallow some large items whole. Adult frogs are active predators, mainly of insects, but occasionally eating small lizards, snakes, birds, mice and other frogs. The change in diet after metamorphosis from tadpole to adult is reflected in radical changes in the gut; the long, coiled alimentary canal of the tadpole, suitable for an herbivorous diet, is much shortened and the horny jaws and tooth rows of the tadpole, specialized for algal grazing, are also lost. As far as is known few species are specialized in their adult food preferences. The *Notaden* species and *Myobatrachus gouldii* are exceptions, living predominantly on ants and termites. Most frogs feed indiscriminately on living insects and their larvae, as well as on other terrestrial arthropods. Sight is the most important sense eliciting feeding behaviour and only moving prey are detected, except in *Bufo marinus*. After recognition of the prey, the frog stalks to within striking distance; long-legged species may leap on to the victim from some distance away. Small prey are captured using the muscular tongue, which, unlike all other vertebrates, is attached at the front of the mouth and is free at the rear. This enables it to be flipped out with great speed and accuracy and the victim becomes attached to the sticky mucus of the tongue. Some frogs have small teeth on the margins of the upper jaws. These are used only for holding prey, which is then swallowed whole. Swallowing is assisted by a muscular retraction of the eyes, so that they bulge down into the mouth (buccal) cavity. Hands may also be used to push large food items into the mouth. *Cyclorana platycephala* and *Litoria dahlii* are two of the few species that can capture food items under water.

In addition to insects and other arthropods, some frogs prey on smaller species and juveniles of their own kind. Notable for this among Australian species are *Cyclorana australis, Litoria aurea* and *Litoria albo-guttata*. Some frogs will also eat small mammals and reptiles, and a large Cane Toad (*Bufo marinus*) which once dwelt at Taronga Zoological Park in Sydney often ate three adult mice, consuming this meal in less than a minute.

After swallowing, the food traverses a short gullet to the stomach where digestion begins. Final stages of digestion are accomplished in the small intestine and absorption of nutrients then occurs. The hard, indigestible parts of an insect, exoskeleton, mouth-parts etc., are compacted in the rectum and pass out with other faeces through the cloaca.

The food requirements of frogs vary with temperature and the stage of the breeding cycle. At low temperatures frogs feed only infrequently, whereas reproductive activity or preparation for long periods of dormancy require larger food intakes.

Captive frogs, with food *ad libitum* may grow abnormally large, such as a *Litoria rothii* which grew to 104 mm snout-vent length, more than twice the previously known maximum size — on a diet of mice.

Respiration and circulation

Metabolic processes by which energy is made available for muscular or chemical work require an uptake of oxygen and release of carbon dioxide. The blood system transports oxygen and carbon dioxide between the respiratory surfaces and the body tissues — the sites of metabolism. It is therefore convenient to consider the respiratory system and the blood system together.

Oxygen uptake and carbon dioxide release occur by diffusion across the moist surfaces of the lungs, skin, and mouth cavity. The lungs are ventilated by intermittent gulping movements. The mouth cavity is ventilated by regular raising and lowering of the throat, drawing air in and out of the mouth through the nostrils (nares). The skin is not actively ventilated. All respiratory surfaces are perfused by blood-filled capillaries, and circulation of the blood is achieved by the pumping action of the muscular heart. In tadpoles, respiration occurs via gills, as well as across the skin, until they are lost at metamorphosis.

The blood system has many functions apart from the transport of respiratory gases. It serves as a conduction system for food materials and hormones, while excess nitrogen, a by-product of protein metabolism, is removed from the blood as it traverses the kidneys and is excreted as urea in the urine.

Predation and defence

Although they are effective predators, frogs themselves fall prey to a great variety of animals including other species of frogs. Predation is most severe during egg and larval stages when both eggs and tadpoles are eaten by fish, snakes, many wading birds and invertebrates. The tadpoles of *Lechriodus fletcheri* and some other species will eat eggs of their own and other species. The introduction of the Mosquito Fish *Gambusia affinis* to Australia's fresh waters for mosquito control has resulted in severe predation on eggs and larvae of native frogs. Tadpoles have few defences against enemies. They may escape by swimming or burrowing into the soft bottom of ponds among plant debris. The tadpoles of some species have chemical secretions which make them distasteful.

Adult frogs are savoured as food by snakes, birds, water rats, and fish (Plate 6). Even large spiders and preying mantids are known to prey on the smaller frogs. A frog's best method of defence is rapid escape by leaping, often into the water. Camouflage is an important defence mechanism and by remaining still many cryptic species escape detection. Chemical defences are also important and the granular glands in the skin of some species exude noxious chemicals when touched. *Notaden* species exude copious amounts of a sticky secretion when molested. Toxins produced by Australian frogs are not known to be dangerous to the human skin but are often extremely irritating on contact with the mouth or eyes. Secretions from the parotoid glands of *Bufo marinus* are reported to be fatal to some fish, snakes, monitor lizards, birds, and mammals. In addition, species which are not poisoned by the toad or their tadpoles often find them distasteful and will not eat them.

Despite the chemical defences in their skins, some may be attractive to mosquitoes (Plate 7)

An important factor in survival is the secretive nature of most frogs by day. Many species burrow and others hide under logs and stones. Tree frogs often hide among green vegetation or under the loose bark of trees. This behaviour facilitates water retention during the dry daylight hours, but is also of great importance in avoiding predators.

Further defensive mechanisms are seen in the larger toad-like species which, when molested, inflate their lungs and become more difficult to grasp. Some species of the toadlet *Pseudophryne* adopt a characteristic head-down posture and often play dead, remaining motionless for several minutes.

Temperature relations

Anurans have no means of generating and conserving body heat, and are thus known as ectotherms (= "cold-blooded"). Consequently, they are usually at the same temperature as the environment. This characteristic is one reason why environmental

PLATE 5. An example of camouflage: *Litoria nannotis* on a lichen-covered tree trunk; Mt Lewis, northern Queensland.

PLATE 6. *Litoria latopalmata*. Predation by the Yellow-faced Whip Snake *Demansia psammophis*; Boowinda Gorge, Carnarvon National Park.

PLATE 7. *Litoria infrafrenata* under attack by mosquitoes.

PLATE 8. *Litoria flavipunctata* bask in the sunshine in a pond near Guyra, New South Wales.

PLATE 9. Ventral view of a gravid female *Heleioporus psammophilus*. Large yellow eggs are seen through the body wall; south-western Western Australia *(Specimen courtesy Grahame Webb)*.

PLATE 10. Male *Litoria fallax* calling at Killibakh, New South Wales. The distended vocal sac acts as a resonating chamber *(Photo by Glenn Hoye)*.

temperatures play an important role in determining geographic distribution. Because of evaporation of water from the moist skin, a frog's body temperature may be even lower than that of the surrounding air.

The only control frogs can exert over their body temperature is by behavioural means, selecting places where the temperature is neither too hot nor too cold. Basking in the sun has been observed in several species, including *Litoria aurea* and its close relatives (Plate 8) and *Cyclorana,* and measurements show that the body temperature may be raised several degrees. It is likely that basking occurs only when there is unrestricted access to water. Behavioural selection of suitable environmental temperatures is frequently seen in tadpoles. In a large pond, for example, it is common to see all the tadpoles clustered together in one shallow area where, under the influence of the sun, the temperature may be several degrees warmer than elsewhere in the pond.

Because body temperature determines the rates of metabolism, digestion, cell division and almost all bodily functions, the dependence of body temperature on environmental temperature has great significance in the life of frogs. Warm or cold weather profoundly modifies the time taken for eggs to hatch and for tadpoles to develop. Thus, seasonal conditions may influence the success of a whole season's reproductive activity in a particular locality. In a cold season, for example, metamorphosis of tadpoles may not occur before the winter sets in. Some tadpoles, such as those of *Limnodynastes peronii* and *Mixophyes schevilli* are able to overwinter.

Reproduction

Reproduction in anurans, as in all sexually reproducing animals, involves the union of sperm and ovum to form a zygote which will divide and develop into the new individual. This union (fertilization) usually occurs in water. Complex physiological and behavioural events lead to successful fertilization. Reproductive products (sperm and ova) must be produced synchronously by males and females, and reproductive behaviour coordinated so that fertilization occurs. All of these events must be synchronized in such a way that breeding can occur when conditions are suitable. Many species breed seasonally and apparently use temperature, day length, moisture and atmospheric pressure cues to time reproduction. The need for free water for fertilization in most species (but not in microhylids) accounts for rainfall serving as the final trigger of reproductive behaviour. However, many species live in areas where rainfall is a sporadic rather than a seasonal event. In these cases, such as in some Australian desert species, breeding is triggered at short notice by heavy rainfall without the necessity for other seasonal cues. This enables the desert species to take advantage of temporary bodies of water. Hatching and larval development are accelerated in these species.

Just prior to spawning, eggs are shed from the ovaries and find their way into the openings of the paired oviducts through which they pass to the outside (Plate 9). Meanwhile, frogs which may have dispersed widely seeking shelter and food are converging on the pond. Mating calls are of great importance in guiding the frogs to ponds. Early male arrivals call with characteristic voice (Plate 10) and this probably encourages others of both sexes to migrate to the pond. Eventually, large numbers of frogs, often of several different species, may be present. At breeding, the male grasps the female from behind (this is termed amplexus) and the two frogs float and swim together in the water. The male may grasp the female over or under the arms (axillary clasp, Plate 11), or around the waist (inguinal clasp, Plate 12), depending on the family. The nuptial pads on the thumbs of breeding males, produced in response to circulating hormones, facilitate this grasp. At amplexus, males often knead the females body to stimulate the release of eggs. As eggs are shed, sperm are released over them and few eggs fail to become fertilized. The protective jelly coat covering the egg then absorbs water and swells. In some genera (e.g., *Limnodynastes*), the female beats the

PLATE 11. Male and female *Litoria brevipalmata* in amplexus, showing axillary clasping; Ourimbah, New South Wales.

PLATE 12. A pair of *Taudactylus acutirostris* in amplexus, showing inguinal clasping; near Kuranda, northern Queensland.

PLATE 13. Floating foam egg mass typical of *Limnodynastes*, in this case, *L. peronii*. The male may remain at the site for several days; Frenchs Forest, New South Wales.

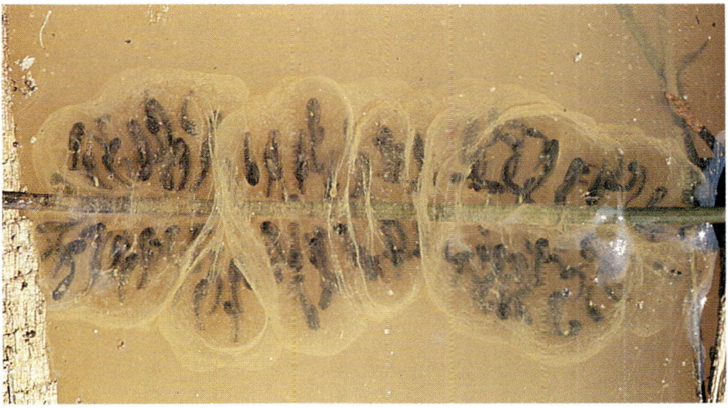

PLATE 14. Jelly egg mass of *Litoria ewingii*, attached to submerged vegetation. Embryos are well advanced; Cann River, Victoria.

PLATE 15. Jelly egg mass of unidentified Hylid; near Guyra, New South Wales.

PLATE 16. Eggs in tangled jelly strings, *Bufo marinus*; Gayndah, Queensland.

water with her front feet, capturing air bubbles which are driven under her body and trapped in mucus to produce the familiar foam nests (Plate 13) in which the fertilized eggs become ensconced. Directing air into the mucus is aided by the female's flanged fingers. In most *Litoria*, the eggs are attached in loose masses, without foam, to underwater vegetation (Plates 14, 15), but some (such as *Litoria coplandi*) drop single eggs to the pond floor. Eggs of the Cane Toad *Bufo marinus* and species of *Neobatrachus* are laid in strings of jelly (Plate 16). Hatching of the tadpoles may take one or several days and larval life varies from one week to many months and, depending upon the temperature, timing may be very variable within a species. A series of photographs illustrating the life cycle of *Limnodynastes tasmaniensis*, is seen in Plates 17–24.

Many patterns of reproduction are not typical. In the mountain-dwelling genera *Kyarranus* and *Philoria*, for example, development occurs within a mucous mass in a burrow away from free water. The free-living tadpole stage is suppressed and metamorphosis coincides with hatching. *Cophixalus* and *Sphenophryne* lay a chain of eggs on land resembling a string of beads, and *Mixophyes* lay their eggs in leaf litter beside a stream, the embryos being washed into the water where tadpoles complete development. *Arenophryne* and *Myobatrachus* lay their eggs deep in moist soil, whereas *Assa* and *Rheobatrachus* carry their tadpoles about with them.

Water relations

The dependence of frogs on water, because of their moist skin and external fertilization, has been discussed already. The freely permeable skin exposes frogs to desiccation when on land and to osmotic influx of water when they are in the water. Where there is an influx of water the excess is removed by the kidneys. In order to regain or conserve water, frogs can absorb water from contact with damp surfaces and furthermore have the ability to reabsorb water from urine stored in the bladder. These processes are regulated hormonally.

PLATE 17. *Limnodynastes tasmaniensis* — Male calling . . .

PLATE 18. . . . A female approaches, responding to his call . . .

PLATE 19. . . . Amplexus, egg mass at an early stage.

PLATE 20. At a later stage, the male is now completely hidden in foam.

PLATE 21. The completed egg mass. Individual eggs are visible as small dark spheres.

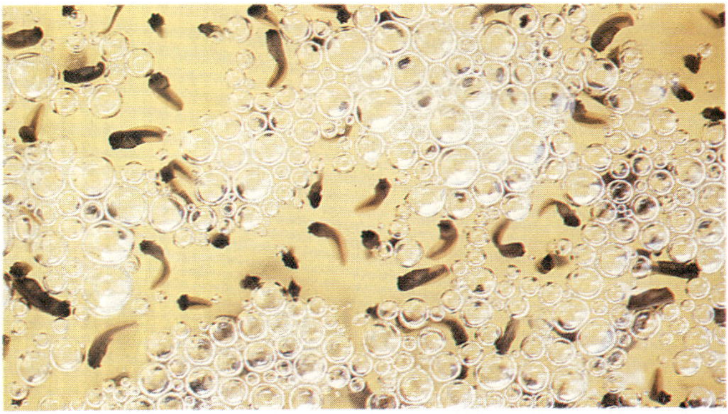

PLATE 22. The foam begins to disintegrate in later stages of embryonic development.

PLATE 23. Tadpole with hind limbs beginning to develop.

PLATE 24. Late stage of metamorphosis, prior to resorption of the tail.

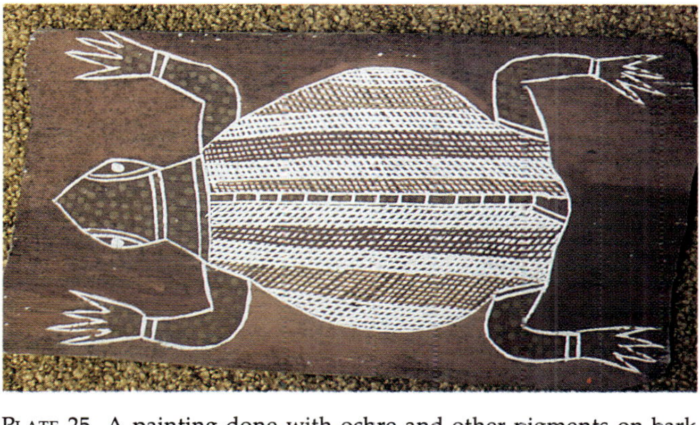

PLATE 25. A painting done with ochre and other pigments on bark by Peter Numberaitj, a Gunwinggu man from Mormega, Arnhem Land, in the Northern Territory.

Considering their water requirements, the extent to which frogs have colonized some remarkably dry regions is surprising. Survival in such areas depends mainly on behavioural patterns which avoid exposure to the arid conditions. These include burrowing habits and long periods of dormancy. The most extreme adaptations are seen in those frogs which secrete a cocoon of shed skin layers around themselves to retard water loss during dormancy. In perhaps no other country is the potential for research into arid adaptations of frogs so great as it is in Australia.

Frogs and humans

Frogs occupy an important position in the food chain, forming a major part of the diet of many fish, snakes, birds and mammals. They may have direct economic significance for man in their predation on insect pests. Frogs have a long association with human folklore and are usually represented in an unfavourable light. No witch's potion could be complete without some frog or toad extract, and many a handsome prince was turned into a frog as punishment for some misdeed. A persistent part of folklore is the erroneous belief that handling frogs causes the development of warts. Despite the recurrent idea that frogs are loathsome, slimy animals, they are actually the least offensive of all the vertebrates. Only a few bite, and any noxious secretions are used exclusively for defence. Part of their appeal derives from their bright and often vivid coloration, particularly in tropical species, which may be seen as a warning of their noxious chemical defences.

In some parts of the world, frogs are regarded as a delicacy and the legs of the European Edible Frog *Rana esculenta* and the North American Bullfrog *Rana catesbeiana* are savoured. The latter species is cultured in many countries for the restaurant trade. Both species grow to a large size and this has probably assisted their culinary acceptance. In India, over-collecting of frogs for the European market has led to population explosions of certain insect pests.

Frogs formed a minimal part of the diet of the Australian Aborigines. The Water-holding Frog *Cyclorana platycephala* and other burrowing species were sought as a handy source of drinking water when other supplies failed. By tapping with a stick or stamping on the surface of a dried-out creek bed or claypan, some Aborigines were able to locate frogs holed up for the dry season.

Frogs are also prominent in Aboriginal folklore (Plate 25). In some Queensland tribes it was customary for corroboree singers to "train" on frogs. The singer would keep a certain type of brown frog in a container of water until the water was sufficiently "rich", then drink the water and the frog. This was thought to benefit the voice.

In the future, frogs may be of more direct significance to humans since some of the glandular secretions of the skin have been found to have important medicinal properties. Some of the active substances in these secretions have been isolated and synthesized as a source of new drugs for the treatment of a variety of infections and other conditions.

One issue of particular concern is that there are world-wide reports about declines of frog populations. Some species seem to be unaffected, yet others are in marked decline. There seem to be no particular correlations with habitat or taxonomic affinity. Some populations are declining in agriculturally-affected lands, prompting concern about the use of insecticides, fertilizers and other chemicals. However, many species in apparently pristine areas are also declining, which may imply a global rather than a local cause. But why, then, are not all species affected? The problem is a difficult one to come to grips with because frog populations may vary very considerably in density from time to time anyway, as good and bad seasons come and go, and because it is very difficult to make an objective estimate of population densities of frogs. Further, there are insufficient baseline data, so much of the concern is based on scientifically unsubstantiated information rather than quantitative data, strong though the evidence may be. In Australia, the two species of *Rheobatrachus* may be extinct, only three of the six species of *Taudactylus* have been seen in recent years and there is concern about all three eastern Australian species in the *Litoria aurea* complex (two of which were previously widespread and abundant), the *Mixophyes* complex as well as others. Some references to information or opinions about the status of individual species will be made throughout the text. Readers are encouraged to collect and communicate information to the Endangered Species Unit of the Australian Nature Conservation Agency, or to their respective state museum.

A part of every zoologist's training is a study of the anatomy of frogs by dissecting the largest easily available "local" species. The use of frogs as a source of material for experiments in nerve and muscle physiology, developmental biology, membrane biophysics, and hormone physiology has contributed enormously to our knowledge in these fields. Nevertheless, relatively few zoologists learn to understand frogs as living animals in all their ecological, behavioural, taxonomic, and colourful diversity. We hope that this book may promote interest in some of these lesser-known aspects of the biology of frogs.

Additional reading

Readers are referred to the magnificent and comprehensive "Biology of Amphibians" by Bill Duellman and Linda Trueb (1986) (see Selected Bibliography, p. 390).

A recorded tape of mating calls

A cassette recording of the male advertisement calls of 44 species of southeastern Australian frogs is available from Gordon Grigg, Department of Zoology, The University of Queensland, QLD 4072. Wherever applicable, we have made reference to this recording in the descriptions of species.

Chapter 3

The Australian Frogs

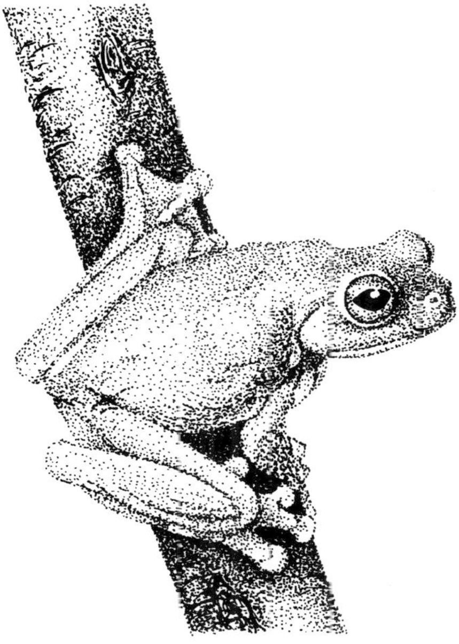

The 204 species of Australian frogs currently recognized are grouped into 28 genera*, representing five separate families. These families, none of which is restricted to Australia, are the Hylidae, Myobatrachidae (also known as Leptodactylidae), Microhylidae, Ranidae and Bufonidae.

Major anatomical features are used to define similarities and differences between families of frogs. For example, the bone structure of the shoulder girdle, characteristics of the vertebral column or limbs, or the arrangements of muscle groups may be considered by taxonomists.

Of the five families found in Australia, two are represented by single species. The Bufonidae, a major family in many countries, is represented only by the introduced Cane Toad *Bufo marinus*. Bufonids are the true toads, so it can be said that Australia has no native toads. Another of the world's most widespread families, the Ranidae, is represented by only one species, *Rana daemeli*, in far north Queensland and the Northern Territory. This species is more widely distributed in New Guinea where there are also many other species of *Rana*.

The Microhylidae is diverse in New Guinea, but in Australia is confined principally to the rainforests of Cape York Peninsula of Queensland, where 15 species are known so far. The sixteenth known species is *Sphenophryne adelphe* of the Northern Territory where it has been found on mown lawns as well as at less disturbed sites.

The dominant Australian families are the Hylidae with three genera and 78 species and the Myobatrachidae (23 genera and 96 species). Although popularly called Tree Frogs, hylids include numerous species that live upon the ground and a few that are burrowers (fossorial). In contrast, the Myobatrachidae is composed principally of ground dwelling and burrowing species, whilst climbing myobatrachids are rare.

---

*Now 29; the genus *Bryobatrachus* has been reported recently from Tasmania. (Note added in proof.)

Unfortunately, taxonomists frequently disagree on familial classification and there is much debate about the status of the Australian "Leptodactylidae-like" frogs. Some authors would prefer to classify them with species from elsewhere in the Leptodactylidae. Others see sufficient differences in the Australian "Leptodactylids" to warrant a separate family, the Myobatrachidae. We will use Myobatrachidae as the familiar name, to be consistent with the recently published *Fauna of Australia* Volume 2A, Amphibia and Reptiles.

Guide to the characteristics of families

As indicated in Chapter 1, evolutionary convergences, whereby similar frog "models" have evolved more than once, make it extremely difficult to distinguish externally between similar models of different families. For example burrowing hylid frogs (genus *Cyclorana*) may appear morphologically very similar to myobatrachid burrowers such as *Neobatrachus*. To confuse matters even further, the characters that distinguish the families are internal ones. It follows that correct identification of genera and species is more likely to be made from comparison with photographs and perusal of distribution maps, than by stepwise progression through the keys. Nevertheless the keys highlight the subtle differences between genera, and between species in each genus. Use of the keys will become easier with practice, until a stage is reached where they outlive their usefulness to the reader.

Family Hylidae (p. 43): Most hylid frogs have expanded and flattened discs on the tips of their fingers and toes. On the undersurface of the discs there is a slight indentation at the margin (the circum-marginal groove). Burrowing species lack expanded discs on the digits and possess an unpigmented spade-like outer metatarsal tubercle. Almost all Australian hylids lay their eggs in clumps in water (never in chains or foam nests). Some members of the hylid genus *Cyclorana* are easy to misidentify as Myobatrachidae.

Family Myobatrachidae (p. 172): Myobatrachids usually have cylindrical fingers lacking expanded and flattened discs. Species which have discs lack the circum-marginal groove. Burrowing species have a spade-like inner metatarsal tubercle that is unpigmented, edged with black or entirely black. Myobatrachids have diverse forms of eggs which can be laid in water or on land, in foam nests, chains or in small clumps.

Family Microhylidae (p. 357): The fingers of microhylids are cylindrical or disced. The disced species have a truncated rather than a circular disc, as though the outer edge has been clipped off with a pair of curved nail scissors. Fingers and toes are unwebbed. Eggs are large, unpigmented, and are probably laid on land. This implies that embryonic development takes place entirely within the egg capsule, with no free-swimming tadpole stage.

Family Ranidae (p. 376): This streamlined, agile frog is the only species in which the male has paired vocal sacs emerging from the angle of the jaws on each side. Toes are fully webbed and there is a dorso-lateral fold from the eye to the hind limb.

Family Bufonidae (p. 379): The dry, warty skin is not sufficient to distinguish *Bufo marinus,* but a simple character not shared by any native species is the curved ridge of bone behind each eye.

Notes on the use of keys

The keys in this book have been designed primarily to assist in the identification of living adult specimens.

In the keys, each numbered section consists of two parts. To identify a specimen, once you have decided which key is appropriate, check which part of each couplet best fits the specimen and proceed to the number shown. Repeat this process of elimination until a name is reached. Remember that the identification should then be confirmed by reference to the description and the photograph. The section in each description relating to similar species will give further assistance.

The keys have been designed primarily to assist in the identification of living adult specimens. Juveniles are often extremely difficult to identify because they commonly show characteristics which differ from those of adults.

Groups and Complexes

In some cases, keys are provided to what we have called "Groups" or "Complexes" of species. Our idea of a group is a subset of species, within a genus, which have distinct physical similarities to each other. A complex is a more loose-knit assemblage. Both are terms of convenience for the purpose of identification. A close relationship is not implied.

Note about tadpoles

Wherever possible information has been provided to assist in the identification of tadpoles. However, it should be remembered that features such as colour may vary depending on the tadpole's habitat because many of them change colour according to the substrate upon which they are resting. Also be aware that, in the wild, tadpole size and the length of the tadpole "stage" may vary considerably according to water temperature, nutrition and other factors.

FAMILY HYLIDAE

Traditionally, members of the family Hylidae are known as tree frogs. The majority of hylids have the ends of the fingers and toes expanded into flattened discs, the undersurfaces of which secrete a sticky mucus-like substance as an aid to climbing. Australia has three genera of Hylidae, *Litoria, Nyctimystes* and *Cyclorana*. A number of *Litoria* species lack expanded discs and are sometimes referred to collectively as "ground hylids". Species of *Cyclorana* also lack these discs. This genus comprises burrowing frogs which, in their body form, closely resemble burrowing frogs of other families. *Cyclorana* species are therefore more likely to be confused with Myobatrachidae than with other Hylidae, so care is needed.

Members of the Hylidae occur on most continents, but the major centres of diversity are South and Central America, New Guinea and Australia. In Australia they are found throughout, with the largest number of species in the north and the east.

Key to the Australian genera of Hylidae

1a. Pupil forms a vertical slit when constricted; lower eyelid bears a pattern of metallic veins .... *Nyctimystes* (p. 44)
b. Pupil forms a horizontal slit or a diamond shape when constricted; lower eyelid lacks pigmentation 2

2a. Hands and feet with or without terminal discs; inner metatarsal tubercle not shovel-shaped (except in *L. alboguttata*) .... *Litoria* (p. 47)
b. Hands and feet lack terminal discs; inner metatarsal tubercle shovel shaped .... *Cyclorana* (p. 153) .... and *L. alboguttata* (p. 142)

Genus *Nyctimystes* Stejneger

Nyctimystes is principally a Papuan genus currently including 22 species. It ranges from the island of Halmahera west of New Guinea, through the New Guinea mainland to the islands that lie to the south-east. Formerly thought to include three species, it is now thought that there is just a single species in north-east Queensland.

The distinguishing features of *Nyctimystes* are the shape of the pupil when constricted, forming a vertical slit, and a pattern of markings upon the lower eyelid. The nature and orientation of these markings vary from species to species and, in some cases, the identification of the species can be made from the eyelid pattern alone. *Nyctimystes* species tend to have large and protuberant eyes, and the eyelid pigmentation presumably acts as a camouflage when the frogs are at rest during the day.

Nyctimystes dayi

(Günther)

Australian Lace-lid

This is a beautiful little frog, highly variable in colour and pattern. Often, the finely shagreened dorsum is a rich tawny brown and many specimens have a few rosette-like white marks scattered irregularly over the back. The ventral surface is creamy white. The snout is flattened, short, and rounded. The eyes are very large with a dark brown iris. The lower eyelid is seen

Nyctimystes dayi, Pilgrim Sands, north Queensland *(Photo by Mike Mahony).*

Nyctimystes dayi, Mulgrave River, north Queensland *(Photo by Gordon Grigg).*

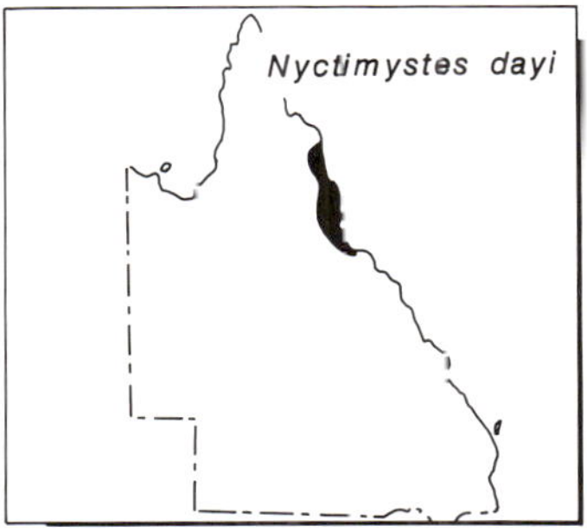

Nyctimystes dayi, note venation on lower eyelid. Mulgrave River, north Queensland *(Photo by Gordon Grigg).*

Nyctimystes dayi, unpigmented eggs laid under a rock. Mt Spec, north Queensland *(Photo by Steve Richards).*

best when the frog adopts a cryptic posture in which the membrane is drawn up over the eye. The tympanum is small and its upper margin is hidden beneath a skin fold that extends from the back of the eye to the insertion of the forearm. The fingers have lateral skin flanges and are partly webbed. The toes are fully webbed. Another interesting and unusual feature is the slight scalloping of the posterior edge of the forearm and tarsus. Similar but more conspicuous scalloping is seen in *Litoria eucnemis*.

SIZE: Males 30–42 mm; females 45–55 mm.

HABITAT: Swiftly flowing rocky streams in the rainforest of the wet tropics of north Queensland.

ADVERTISEMENT CALL: A short, sharp "ee" repeated each 5–6 seconds.

BREEDING AND LARVAE: Breeding takes place through spring and summer. Large, unpigmented eggs are laid in clumps of approximately 100 attached to the undersurface of rocks in the fast flowing creeks. Tadpoles in their early stages are pale yellow, but turn brown in later stages. They have flattened bodies, strong tail muscles, and an unusual mouth which is large, sucker-shaped and located on the undersurface of the head. Tadpoles take a minimum of 3–4 months to complete their development, but those from spawn laid in late summer may remain as tadpoles until the following summer. The tadpole tooth row formula is $\frac{2}{3}$.

SIMILAR SPECIES: The species most likely to be confused with *N. dayi* are *Litoria eucnemis* and *L. genimaculata* which differ in having better developed skin folds upon the outer edges of the limbs, a horizontal pupil, and in lacking the pattern of pigment upon the lower eyelid.

Genus *Litoria* Tschudi

Litoria is one of the most diverse genera of frogs in the world, including some species that climb and live above the ground, others that exist upon the ground, one that burrows and another that skips upon the surface of the water. Perhaps because the physical adaptations to each of these modes of life are so different, there is not a single external feature common to all members of the genus. We have no doubt that further studies will lead to *Litoria* being split into several distinct genera. At present 61 species are recognized in Australia, with many more in New Guinea, eastern Indonesia and the Solomon Islands. Six species are found in both Australia and New Guinea.

The arboreal species of *Litoria* are generally colourful frogs which have flattened heads and bodies and expanded, circular discs on the tips of their fingers and toes. The upper and lower surfaces of the discs are separated by a groove termed the circum-marginal groove. Several of these climbing species have webbing between the fingers and/or flanges of skin upon the edges of the fingers, and all have extensive webbing between the toes. In contrast, the ground-dwelling species have cylindrical heads and bodies and fingers that bear only slightly rounded discs, with neither lateral flanges to the fingers nor webbing between them. Webbing between the toes is less extensive. The sole burrowing species, *L. alboguttata* is equipped with an enlarged inner metatarsal tubercle, and is therefore similar to *Cyclorana* species. *L. meiriana* is a small water-skipping species and has oval discs on fingers and toe discs, and considerable webbing between the toes.

The Australian *Litoria* species usually lay their eggs in water, except the poorly known *L. longirostris* of the northern Queensland rainforests which deposits spawn on the bases of tree trunks at the edge of water. Steve Richards and Ross Alford have described construction of a circular nest in sandy substrate by a *Litoria lesueuri*-like species. Breeding occurs mostly in spring and summer, although frogs of the *L. ewingii* group tend to breed opportunistically after rain. There is a great diversity in mating calls, ranging from the deep barking of *L. infrafrenata* to the high-pitched buzzing of *L. microbelos*. Spawn is usually laid in clumps attached to aquatic vegetation or rocks. Exceptions are *L. coplandi* whose eggs are scattered singly and sink to the floor and *L. caerulea* whose eggs are laid in a single-layered floating mat. The tadpoles are fully aquatic in all species and, in *L. aurea* and *L. caerulea*, may be up to 100 mm long.

Most of the species can be placed conveniently in species groups or "complexes". We have arranged the following descriptions in these categories in order to facilitate easier identification and comparison.

Key to groups and complexes of *Litoria*

1a. A series of irregular lappets of skin along the trailing edges of the arms and legs ...... *Litoria genimaculata* group (p. 49)
 b. Trailing edges or arms and legs lacking lappets 2

2a. Legs extremely short: tibia approximately one-third of the length of the body .... *Litoria rubella* group (p. 53)
 b. Legs moderate to long: tibia greater than one-third of the length of the body 3

3a. Black and yellow/orange markings on back of thighs 4
 b. Back of thighs yellow/orange or other colour, but lacking black markings 6

4a. Patches of black behind tympanum .... *Litoria lesueuri* complex (p. 59)
 b. No black patches behind tympanum 5

5a. Pattern of markings on back and sides extending from front to back .... *Litoria ewingii* complex (p. 82)
 b. Pattern of markings on back and sides not arranged in a longitudinal fashion .... *Litoria peronii* group (p. 92)

6a. Releases large amount of mucus when handled .... *Litoria aurea* complex (p. 99)
 b. Does not release large amount of mucus when handled 7

7a. Colour of back predominantly green .... "Green Tree Frogs" (p. 108)
 b. Colour of back not predominantly green 8

8a. Confined to rocky creek beds in wet tropics of north Queensland; male nuptial pad with spines; eggs unpigmented .... "Torrent Frogs" (p. 135)
 b. Not confined to north Queensland; male nuptial pad lacking spines; eggs pigmented .... "Other species" (p. 141)

The *Litoria genimaculata* Group

The *Litoria genimaculata* species group comprises that species and *L. eucnemis*. They are very spectacular frogs of the north Queensland rainforests. Both species were described from specimens collected in New Guinea in the last century. There has been considerable controversy about the identity of two species in Australia that resemble them. The Australian populations given these names are distinguished by differences in call, and there is no doubt that two species are involved. Until their calls are compared with the two populations in New Guinea there will remain uncertainty about whether the correct names are being applied to them.

Key to the *Litoria genimaculata* Group

1a. Advertisement call a series of soft ticks; fingers not fully webbed .... *Litoria genimaculata* (p. 50)
b. Advertisement call a series of short growls; fingers fully webbed .... *Litoria eucnemis* (p. 53)

Litoria genimaculata

(Horst)

New Guinea Tree Frog

The dorsal coloration is variable but commonly rich brown with a broad, dark tri-lobed or bi-lobed mid-dorsal band. Irregularly dispersed over the head, back, and limbs are patches of metallic green and copper shades. There are often distinct transverse bars across

Litoria genimaculata, Cairns, north Queensland *(Photo by Glenn Hoye).*

Litoria eucnemis, McIlwraith Range, north Queensland *(Photo by Steve Richards)*.

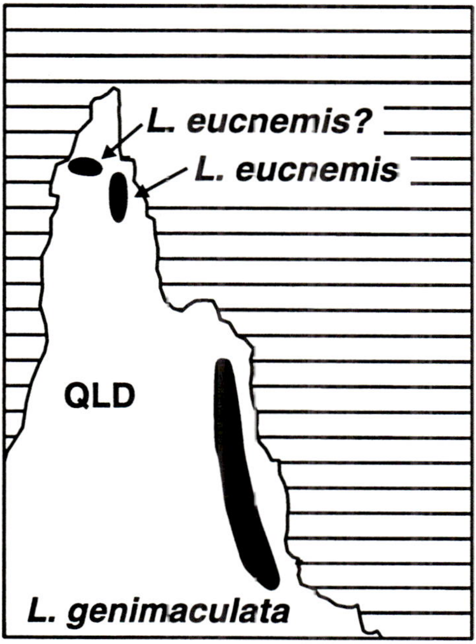

the limbs. These remarkable colours are subdued by bright light. The skin on the back may be smooth or slightly roughened. The head is broad and flattened and the nostrils spaced closely together. The eyes are prominent. A distinctive feature is the green coloration of the upper half of the iris. The canthus rostralis is strongly curved. There is an irregular, scalloped margin of skin along the outer surface of the foot and a small, triangular flap on the toe. Vomerine teeth are between the choanae. Both hands and feet have large discs and well developed webbing.

SIZE: Males 30–41 mm; females 40–52 mm (but see "comment").

HABITAT: Associated closely with rocky creek beds.

ADVERTISEMENT CALL: A series of soft clicks.

BREEDING AND LARVAE: According to the distribution data of Richards *et al.* (1993), the description of the development of *L. eucnemis* by Davies (1989) refers to this species. Oviposition in captivity took place in August. The mass of 843 eggs was embedded in very firm jelly capsules so that the clump retained its shape when picked up. Eggs are dark brown above and cream below and range 2.1–2.6 mm diameter. The tadpole is dark brown and elongate with poorly developed fins. It attains a maximum length of 31 mm and development is completed in a minimum of 63 days. The tadpole tooth formula is $\frac{1\;^{1}\;1}{1\;_{2}\;1}$.

SIMILAR SPECIES: Most likely to be confused with *L. eucnemis* which has more extensive webbing between the fingers (reaching the base of the discs in most specimens) and is a larger species. *Nyctimystes dayi* shares an irregular skin margin to the edge of the foot, but is readily distinguished by its vertical pupil and possession of a pattern of metallic veins upon the lower eyelid.

COMMENT: Richards, McDonald and Ingram (1993) have demonstrated for the first time that two members of this species group are to be found on the Cape York Peninsula of north Queensland, and they distinguish them principally by the differences in advertisement call. Tyler (1968) examined the type and other material and considered *L. eucnemis* to be a larger species than *L. genimaculata* and to have more extensive webbing between the fingers. Steve Richards advises us that in Australia *L. genimaculata* is larger than *L. eucnemis* (the reverse of the trend in New Guinea). It may be best to identify by call.

Litoria eucnemis

(Lönnberg)

Fringed Tree Frog

The description of *L. genimaculata* can be applied equally well to this species except that the finger webbing is not as extensive, reaching no further than half way along the outer finger.

SIZE: Males 37–48 mm; females 51–77 mm.

HABITAT: Closely associated with rocky creek beds.

ADVERTISEMENT CALL: A series of short growls.

BREEDING AND LARVAE: Unknown.

SIMILAR SPECIES: Closely related to *L. genimaculata*. (See p. 50)

***Litoria rubella* Group**

Key

1a. A broad, dark stripe along the middle of the back .... *L. dentata* (p. 53)
 b. No broad, dark stripe along the middle of the back 2

2a. Distinct, dark markings upon the back; advertisement call a buzzing sound .... *L. electrica* (p. 56)
 b. No distinct, dark markings upon the back; advertisement call a series of separate pulses .... *L. rubella* (p. 56)

Litoria dentata

(Keferstein)

Keferstein's Tree Frog

Litoria dentata is an attractive frog, elongate and with dorso-ventral flattening. The dorsum is pale brown or cream with a broad, dark, tri-lobed central band running from the tip of the snout to the cloaca. The sides of the head and body are also dark. The ventral surface is pale and granular, with a yellow vocal sac in breeding males. The fingers and toes have large discs. There is a slight web between the fingers and a well developed web between the toes; only the disc and one phalanx of the fourth toe are free. The tympanum is distinct and dark. The upper half of the iris is a deep wine-red. Vomerine teeth are between or slightly behind the choanae.

SIZE: Males 32–40 mm; females 39–44 mm.

HABITAT: This species is found around large coastal swamps or ponds, either on the ground or in low vegetation. The frogs are very secretive by day, their flattened bodies enabling them to hide under stones or loose bark on trees.

ADVERTISEMENT CALL: A very loud, high-pitched pulsed note which has a duration of nearly one second. It is a very harsh sound and the noise produced by a large chorus is almost painful to the ears. (Tape track 27).

BREEDING AND LARVAE: Males call during heavy spring and summer rain. The calling position is usually on the ground amid grass and twigs, within a few metres of the water. Larvae have a typical hylid body form and mouth structure.

Litoria dentata, Jervis Bay, New South Wales *(Photo by Marion Anstis)*.

Litoria dentata, calling male, Ourimbah, New South Wales (*Photo by Mike Mahony*).

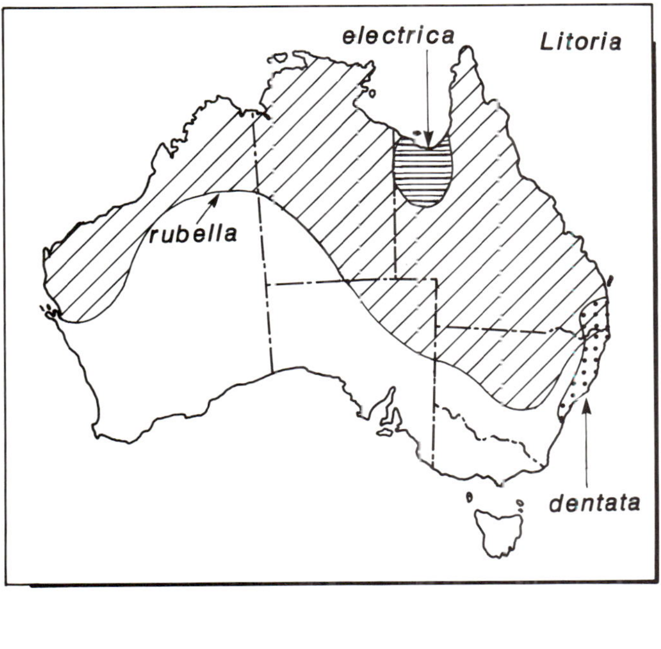

SIMILAR SPECIES: *Litoria capitula* of Irian Jaya, *L. dentata, L. rubella, L. congenita* of New Guinea and *L. electrica* form a species group. The geographic distribution of *L. dentata* does not overlap with any of the other members of the group except *L. rubella* in south-east Queensland. *L. dentata* is distinguished morphologically by the broad, dark, mid-dorsal band.

Litoria electrica

Ingram and Corben

Buzzing Tree Frog

This small and stocky frog has short limbs yet climbs extremely efficiently. The back is brown or has a yellowish tinge and bears darker markings on the back. These may be extensive and in the form of butterfly wings, a chevron or irregular patches. The posterior surfaces of the thighs are marked with brown and pale yellow. The fingers are webbed at the base, and the toes half-webbed and broadly fringed.

SIZE: Males 26–38 mm; females 30–38 mm.

HABITAT: Found in a wide range of situations from the periphery of semi-permanent swamps, to within buildings.

ADVERTISEMENT CALL: Described as similar to the sound of a high voltage electric arc.

BREEDING AND LARVAE: Calling has been recorded in December, around temporary waterholes in eucalypt woodland.

SIMILAR SPECIES: Most similar to *L. rubella* from which it can be distinguished by the dark distinctive shape of the markings upon the back and by its higher, buzzing call.

Litoria rubella

(Gray)

Red Tree Frog

This little frog has a remarkably wide geographic distribution, occupying the northern three-quarters of Australia. The dorsum is pale grey to reddish-brown with a few small black flecks or distinct darker patches.

Litoria electrica, Burketown, Queensland (*Photo by Margaret Davies*).

Litoria rubella, Inglewood, Queensland (*Photo by Gordon Grigg*).

Litoria rubella, different colour phase, Inglewood, Queensland *(Photo by Gordon Grigg).*

A dark band extends along the side of the head and body. The ventral surfaces are white and granular except the throat of breeding males which is a very dark grey. Groin usually lemon yellow. The tympanum is distinct. The limbs are very short and robust, and the fingers and toes have large discs. The fingers have a slight web and the toes are about one-half webbed, the fourth toe having the disc and two phalanges free.

SIZE: Males 28–37 mm; females 34–43 mm.

HABITAT: This frog is one of the few hylids occurring in the arid regions as well as the wet coastal fringe. It is found also in southern New Guinea. By day, *L. rubella* makes use of any available shelter: beneath stones or loose bark, in sheds, buildings or water pipes of houses in very dry areas.

ADVERTISEMENT CALL: A long and distinctly pulsed call slowing and rising towards the end. (Tape track 44).

BREEDING AND LARVAE: Because of its wide geographic distribution, breeding seasons vary according to local conditions. On the northern and eastern coasts, breeding coincides with heavy summer rain. In the drier areas, breeding occurs as the opportunity arises. Males call from the ground within a few metres of the water. The spawn is in the form of a film of golden eggs floating on the surface. The duration of the tadpole stage varies with the water temperature but may be as short as 14 days. The tadpole is pale brown and the tooth formula is $\frac{1\,^{1}\,1}{1\,_{2}\,1}$.

SIMILAR SPECIES: *Litoria electrica* is the most closely related species and can be distinguished by its butterfly markings on the back and buzzing call. *Litoria electrica* is confined to the eastern foot of the Gulf of Carpentaria.

The *Litoria lesueuri* Complex

A large number of species of *Litoria* having similar general appearance and ground-dwelling habits can be grouped conveniently as the *L. lesueuri* complex. Associated with their adoption of a ground-dwelling mode of life, there has been a marked reduction in the size of discs on the fingers and toes and an increase in leg length, enabling them to leap very effectively. Identification of species within this complex can pose considerable problems as many new species have been described in recent years and experience is required for positive identification in some cases. The complex can be divided into two groups on the basis of mating call and general morphology. The *L. lesueuri* group has a soft, almost inaudible mating call (this group lacks vocal sacs) whereas the *L. latopalmata* group has a call consisting of rapidly repeated quacking notes.

The species of the complex are: *L. lesueuri* group (soft chuckling call): *L. lesueuri, L. booroolongensis. L. latopalmata* group (repeated quacking call): *L. brevipalmata, L. latopalmata, L. tornieri, L. watjulumensis, L. inermis, L. nigrofrenata, L. pallida, L. personata, L. freycineti* and *L. nasuta*.

Key to the *Litoria lesueuri* Complex
(including both groups)

1a. Black stripe on side of head passing over tympanum 2
 b. Black stripe on side of head full width of tympanum 4

2a. Webbing between toes vestigial. Green and black markings in concealed areas of groin and armpits ..... *L. brevipalmata* (p. 62)
 b. Webbing between toes extensive. Concealed areas not green 3

3a. Head-stripe indistinct or absent. Dorsum usually warty. Webbing reaches disc of fifth toe .. *L. booroolongensis* (p. 61)
b. Head-stripe very distinct. Dorsum smooth. Disc and half to one phalanx of fifth toe free from web .. *L. lesueuri* (p. 68)

4a. Dark stripe on side of head continuing as an uninterrupted band to the flanks 5
b. Dark stripe on side of head interrupted, and extending to forearm or only slightly beyond 6

5a. Toes fully webbed. Confined to northern Western Australia, the Northern Territory and extreme western Queensland .... *L. watjulumensis* (p. 79)
b. One half of the penultimate phalanx of the fifth toe free of webbing. Confined to Cape York Peninsula of Queensland .... *L. nigrofrenata* (p. 74)

6a. Back smooth or irregularly warty 7
b. Back with warts in longitudinal rows or with longitudinal folds of skin 11

7a. Back with irregularly arranged warts. Stripe on side of head poorly defined .... *L. inermis* (p. 66)
b. Back smooth. Stripe on side of head clearly defined 8

8a. Finger discs wider than finger 9
b. Finger discs not wider than finger 10

9a. Head stripe disrupted behind tympanum; confined to Queensland, New South Wales and South Australia .... *L. latopalmata* (p. 67)
b. Head stripe not disrupted behind tympanum; confined to Arnhem Land, Northern Territory .... *L. personata* (p. 77)

10a. Dark stripe on outer edge of tibia continuous .... *L. tornieri* (p. 77)
b. Dark stripe on outer edge of tibia disrupted .... *L. pallida* (p. 75)

11a. Back with distinct longitudinal skin ridges; posterior of thigh yellow with longitudinal black stripes .... *L. nasuta* (p. 71)
b. Back lacking longitudinal skin ridges, but with warts in longitudinal rows; posterior of thighs light brown with large cream spots .... *L. freycineti* (p. 64)

Litoria booroolongensis

(Moore)

Booroolong Frog

The Booroolong Frog is a highland species, ranging from the Queensland border down the Great Dividing Range nearly to the Victorian border. Although specimens were collected early this century, it was not described until 1961. The dorsum may be grey, olive, or reddish-brown with rather indistinct black reticulations. Its texture is usually slightly warty. The throat is smooth and white in females and dark in males. The abdomen is white and finely granular. A faint, thin, black stripe begins at the snout and passes through the eye, curving slightly over the small distinct tympanum to the shoulder. Vomerine teeth of this species and all others in the *lesueuri* complex are small and situated between the choanae. Fingers and toes have well developed discs. The fingers are free from web while the toes are strongly webbed to the discs. The posterior of the thigh is dark brown with a few very small paler spots.

SIZE: Males 36–42 mm, females 48–54 mm.

HABITAT: *Litoria booroolongensis* is nearly always associated with flowing rocky streams on the slopes and tablelands of the Great Dividing Range. It is often found in daylight on rocks by the water's edge.

ADVERTISEMENT CALL: A soft purring "qrk-qrk-qrk" lasting two or three seconds.

BREEDING AND LARVAE: Males begin calling about August, sitting on rocks near or in the water. A large number of small pigmented eggs are deposited among the rocks. The tadpole is similar to that of *L. lesueuri* and has a tooth formula of $\frac{2}{3}$.

SIMILAR SPECIES: *Litoria booroolongensis* is the easiest of the group to identify. Within the complex, only *L. lesueuri* and *L. latopalmata* have geographic distributions that overlap. The rounded snout, indistinct head-stripe, large discs, and greater webbing permit easy identification.

Litoria brevipalmata

Tyler, Watson and Martin
Green-thighed Frog

Litoria brevipalmata is a colourful frog. The dorsal surface of adults is a rich chocolate brown with many scattered black flecks. The ventral surface is white and granular with fine, dark-coloured spots peppering the throat. The vocal sac is yellow in breeding males. The side of the head is attractively marked with a black stripe from the snout through the eye to the shoulder, where it breaks into a series of blotches which continue along the flank. The margin of the upper jaw is white. Armpits and groin are beautifully marked with lime-green and black markings, a coloration which provides the most striking feature of this species. The hind limbs are rather short compared to other ground hylids. Toes have only a small web while fingers are free from web. The discs are moderately developed and vomerine teeth are between the choanae.

SIZE: Males 41–43 mm; females 45–47 mm.

HABITAT: There is still insufficient information on the distribution and habitat requirements of this species, although some new populations have been discovered in the Brisbane and Ipswich areas. In October large numbers have been found after very heavy rain in a flooded paddock bordering rainforest and they are known to breed in open forest also. Breeding has been observed in October and in January, after very heavy rain, in flooded paddocks or temporary waterholes near rainforest or open woodland. Juveniles with a body length of 15–18 mm have been collected in April from wet sclerophyll forest.

ADVERTISEMENT CALL: In late October 1972 the mating call was recorded for the first time. It is a series of short quacking notes which may continue for two minutes. (Tape track 1).

BREEDING AND LARVAE: The spawn is laid in loose clumps among water weed. The larvae are uniform dark brown to black. Tooth formula is $\frac{2}{3}$.

SIMILAR SPECIES: The brilliant colour in the groin and the unwebbed toes make this species easily recognizable.

Litoria booroolongensis, Tumut, New South Wales *(Photo by Gordon Grigg).*

Litoria brevipalmata, Ourimbah, New South Wales *(Photo by Gordon Grigg).*

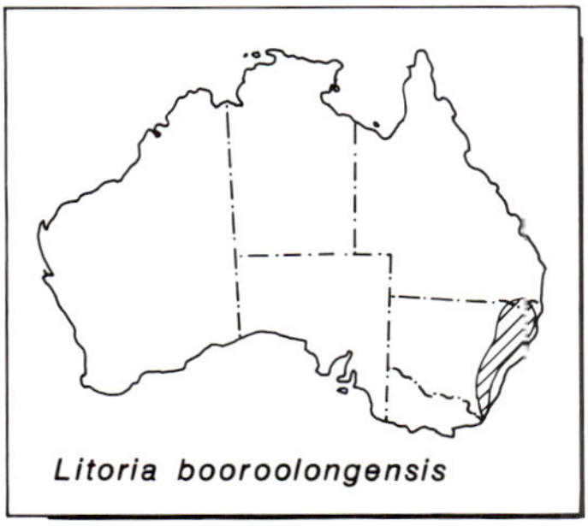

Litoria brevipalmata, pair in amplexus, Ourimbah, New South Wales *(Photo by Gordon Grigg).*

Litoria freycineti

Tschudi

Freycinet's Frog, Wallum Rocket Frog

The dorsum is light brown with a number of large darker warts and skin folds arranged in roughly longitudinal rows. The most consistent feature is a darker triangular patch on the head, often with a light spot in the centre. The ventral surfaces are cream or white and coarsely granular. The throats of breeding males are pigmented on the edge of the lower jaw and a fold is present on the shoulder. The head-stripe in front of the eye is broad, cut by a heavy vertical white bar which sweeps under the eye to the corner of the mouth. Behind the eye, the stripe is the full width of the tympanum (which is outlined in white) and ends at the shoulder, breaking up into a number of blotches. The hind limbs are very long and powerful, the disc and one phalanx of the fifth toe and the disc and two phalanges of the fourth toe are free from webbing. The discs are small. The posterior surface of the thigh is brown with a few large cream spots.

SIZE: Males 34–39 mm, females 39–42 mm.

Litoria freycineti, Helensberg, New South Wales *(Photo by Gordon Grigg).*

Litoria inermis, Dorunda, north Queensland *(Photo by Gordon Grigg).*

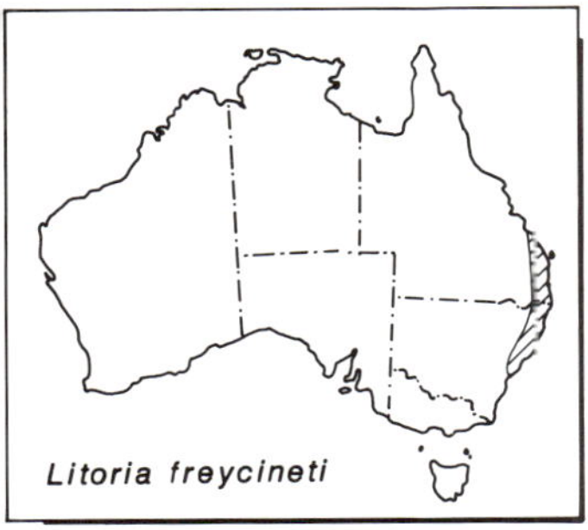

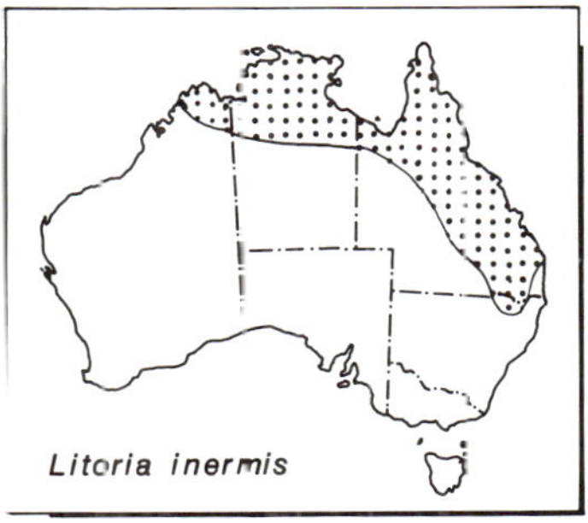

HABITAT: *L. freycineti* occupies a variety of habitats, but is most frequently seen in the vicinity of temporary coastal swamps (the "wallum").

ADVERTISEMENT CALL: Very similar to *L. latopalmata*, although it may be less regular. (Tape track 11).

BREEDING AND LARVAE: Similar to *L. latopalmata*.

SIMILAR SPECIES: See *L. nasuta*.

Litoria inermis

(Peters)

Peters' Frog

The dorsum has numerous small warts and is dark grey or brown with darker patches. The ventral surface is off-white and granular. The snout is not conspicuously pointed and head length is equal to head width. The head-stripe is a wedge from nostril to eye, almost cut by a vertical white bar just in front of the eye. Both these markings are relatively indistinct. The portion behind the eye is almost the full width of the very large tympanum and continues back to the shoulder where it ends abruptly. The hind legs are long, the fifth toe having the disc and one phalanx, and the fourth toe the disc and two phalanges, free from webbing. The discs are very small, barely the width of the digit. The posterior surface of the thigh is yellow with a heavy network of black markings. Fingers are free from web, and the first is considerably longer than the second.

SIZE: Males 24–33 mm; females 30–37 mm.

HABITAT: *Litoria inermis* is found over most of north-western Western Australia, the Northern Territory and north and northeastern Queensland. It is most commonly encountered in open grassland.

ADVERTISEMENT CALL: Males call from exposed sites adjacent to temporary pools from mid-November to mid-March. The call is a long series of short and long notes resembling the rattle of *L. inermis*.

BREEDING AND LARVAE: The spawn is laid in free-floating clumps of 100–330 eggs. Development takes at least 10 weeks. The tadpoles are mottled with brown and have a tooth formula of $\frac{2}{3}$.

SIMILAR SPECIES: From the other ground-dwelling and similar species, *L. inermis* can be distinguished by its warty, dorsal skin, and by the lack of a clearly defined, dark head-stripe.

Litoria latopalmata

Günther

Broad-palmed Frog

This and *L. inermis, L. nigrofrenata, L. pallida, L. personata, L. tornieri* and *L. watjulumensis* are the most difficult to identify in the *lesueuri* complex. They are all characterized by pointed snouts, broad head-stripes, long legs, first finger usually longer than second, and very similar calls and breeding biology.

In *L. latopalmata*, the dorsum is pale grey or brown and may have some indistinct darker blotches and a few small warts. The ventral surface is white and granular. The smooth throat is pigmented on the borders in breeding males and the lower jaw is checked in black and white. The black head-stripe is a thin wedge from the nostril to the eye, nearly cut by a thin vertical white bar just in front of the eye and curving under the eye almost to the corner of the mouth. Behind the eye the stripe is nearly as wide as the tympanum and continues to the shoulder where it breaks into a number of small patches. The legs are fairly long, the posterior of the thigh having a network of irregular patches of black upon a yellow background. There is usually a thin black line or series of spots on the hind edge of the tibia and foot. The toes are long, terminating in very small discs. The disc and half a phalanx of the fifth toe, and the disc and two phalanges of the fourth toe are free from web. The fingers are long, free from web, and have small discs. The first finger is slightly longer than the second. Vomerine teeth are between the choanae.

SIZE: Males 29–39 mm; females 36–42 mm.

HABITAT: *Litoria latopalmata* is found near permanent or semi-permanent pools and billabongs in or near to open forest. At night they may be encountered foraging in leaf litter.

ADVERTISEMENT CALL: All the species in this group have very similar advertisement calls which are very difficult to distinguish, even by electronic analysis. The calls commence slowly and irregularly and build up to a rapid duck-like quacking. (Tape track 2).

BREEDING AND LARVAE: Calling commences in spring when rain coincides with fairly high temperatures, and continues to the end of summer. Males call from the edge of the pond or in damp areas nearby. Loose clumps of small pigmented eggs are deposited among vegetation in the shallows of the pond. The tadpole is of the typical hylid form with two upper and three lower labial tooth rows. The body is light to dark brown. The tail fin has a high crest. Maximum length (body and tail) is 35 mm.

SIMILAR SPECIES: Following the study of Davies, Martin and Watson (1983) it is evident that the only species occurring within the range of *L. latopalmata* and likely to be confused with it is *L. inermis*. The latter may be distinguished by its extremely warty dorsal skin and the lack of a distinct dark stripe upon the side of the head.

Litoria lesueuri

(Duméril and Bibron)

Lesueur's Frog, Stony Creek Frog

It seems that at least two species, large and smaller, have been compounded under this one name: one confined to northeastern Queensland, and the other extending down the coast as far as Victoria. The colour of the dorsum is quite variable, ranging from yellow to dark brown with irregular blotching. The ventral surface is white and granular except for the throat which is smooth and white in females and mottled around the edges with dark brown in breeding males. There is a prominent shoulder fold. A black head-stripe commences at the snout, passes through the eye, and narrows to pass over the tympanum to the shoulder where it breaks into a series of spots on the flank. Vomerine teeth are between the choanae.

Litoria latopalmata, Glen Lyon, Queensland *(Photo by Gordon Grigg)*.

Litoria lesueuri, Wollombi, New South Wales *(Photo by Gordon Grigg)*.

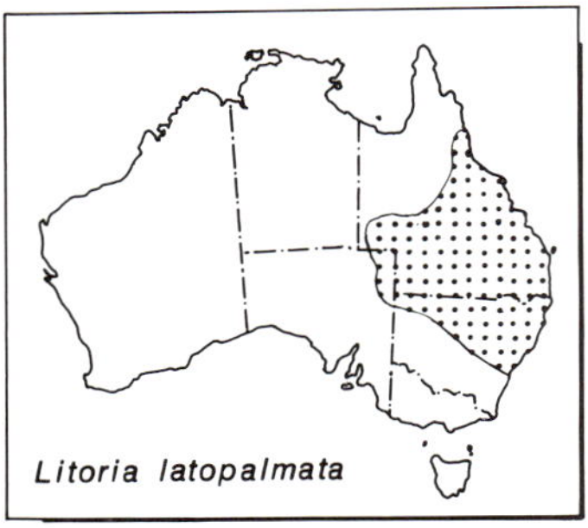

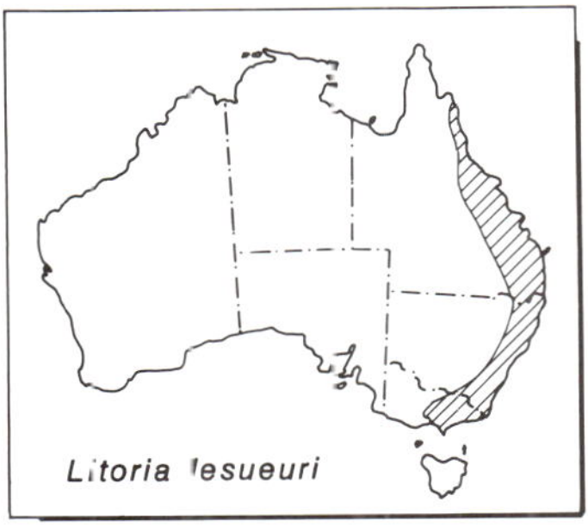

Litoria lesueuri, pair in amplexus, Watagan Mountains, New South Wales *(Photo by Mike Mahony)*.

Litoria lesueuri, nest with single egg mass, Elphingstone Creek, near Ingham, north Queensland *(Photo by Steve Richards)*.

Fingers are free from web and the disc and two phalanges of the fourth toe and the disc and one phalanx of the fifth are free from web. The posterior of the thigh is black with pale blue or yellow markings forming a network or in rosettes.

SIZE: Males 37–43 mm, females 55–63 mm.

HABITAT: Like *L. booroolongensis, L. lesueuri* is usually associated with rocky streams but has been found in a great variety of habitats.

ADVERTISEMENT CALL: Lacking a vocal sac to amplify it, the call is a very soft purring trill repeated for two or three seconds. Males may call during the day. (Tape track 24.)

BREEDING AND LARVAE: Males call from August to May, sitting on the ground or on rocks near the water. Oviposition occurs in quiet reaches of the stream while the amplexing pair lies on the bottom. In the southern population several hundred eggs are deposited in a solid gelatinous clump which adheres to submerged rocks or the bottom sediments. In the northern population eggs are laid in a shallow, circular nest constructed by the frogs. Richards and Alford have described circular sand nests made by this *lesueuri*-like species at the edge of a shallow stream in north Queensland. The larvae have a muscular tail, and narrow fins. The tadpole tooth formula is $\frac{2}{3}$ and the mouth parts are completely surrounded by a thick band of papillae. Length reaches 45 mm.

SIMILAR SPECIES: Due to the nature of the head-stripe, size, and the colour of the thigh, identification should not be difficult.

Litoria nasuta

(Gray)

Rocket Frog

An extremely long-legged, streamlined frog. The colour of the dorsum is in shades of brown with darker longitudinal skin folds or warts usually arranged in two series with a paler area between. The ventral surface is white and granular. Throats of breeding males are pigmented and the edges of the jaws marked in black and white. The head-stripe is very similar to *L. latopalmata* and *L. freycineti*, and the anterior portion is distinctly wedge-shaped. The tympanum has a white border. The hind limbs are extremely long, the posterior of the thigh marked with black longitudinal lines on a yellow background. The disc and one phalanx of the fifth toe and the disc and two phalanges of the fourth toe are free from webbing. The discs are very small. Fingers are free from web, the relative lengths being $3 > 4 > 2 = 1$. The outer metacarpal tubercle is large and slightly nicked.

SIZE: Males 33–45 mm; females 36–55 mm.

HABITAT: *L. nasuta* occurs in the vicinity of coastal streams, ponds and lakes from north of Newcastle, New South Wales, to Cape York Peninsula and New Guinea and westwards to northern Western Australia.

ADVERTISEMENT CALL: A penetrating series of notes "but... but... wick... wick... wick... wick... wick", etc., lasting up to several seconds.

BREEDING AND LARVAE: In northern Australia breeding takes place between November and February. Spawn is laid as a surface film including groups of 50–100 eggs. As in most species, the tadpole life span is linked to the month of spawn deposition, taking one month to complete development if laid late in the breeding season, whereas spawn laid in the early wet season may take almost five months. Tadpoles are mottled with brown and reach a maximum length of 56 mm. The tadpole tooth row formula is $\frac{2}{3}$.

SIMILAR SPECIES: The elongate head and body and extremely long hind limbs render *L. nasuta* one of the most easily recognized species. Confusion is most likely to be within the area of overlap with *L. freycineti* where the two may be distinguished by the texture of the back; narrow skin folds in *L. nasuta* and rows of warts in *L. freycineti*.

Litoria nasuta, Mulgrave River, north Queensland *(Photo by Gordon Grigg).*

Litoria nigrofrenata, Cooktown, Queensland *(Photo by Gordon Grigg).*

Litoria nigrofrenata

Litoria nigrofrenata

(Günther)

Bridle Frog

The dorsum is pale grey to light brown and very smooth. Ventral surfaces are white and granular, except the throat of the male which is lightly pigmented in grey. The head-stripe of *L. nigrofrenata* is quite distinctive, starting as a wedge from snout to eye and cut by a pale vertical bar which curves under the eye to the corner of the mouth. Behind the eye the stripe continues over the tympanum and back to a point just behind the shoulder where it is cut by a thin pale area leaving the last portion as a single patch of black. The hind legs are long (about the same as *L. watjulumensis*) with small discs. The disc and about one half of the phalanx of the fifth toe and the disc and two phalanges of the fourth toe are free from web. The posterior of the thigh is predominantly black with some pale horizontal blotches. The anterior margin of the tibia has a black lens-shaped mark. The fingers are free from web; their relative length is $3 > 4 > 1 > 2$. The posterior edges of the hand and forearm are black.

SIZE: Males 37–42 mm; females 39–46 mm.

HABITAT: This species seems to prefer seasonal creeks where there is ground cover. They also occur around dams and in marshy areas.

ADVERTISEMENT CALL: The call does not appear to have been reported, but the calling site is on the banks of temporary pools.

BREEDING AND LARVAE: Oviposition has been observed from December to February. One clump of spawn included 445 eggs. Tadpoles attain a maximum length of 43 mm. They are characterised by deep, reddish fins and a brown body bearing broad white stripes. The undersurface of the body is copper.

SIMILAR SPECIES: Most likely to be confused with *L. watjulumensis*, *L. nigrofrenata* is distinguished by having less extensive toe webbing and by the nature of the head stripe. However, the two species do not have overlapping geographic distribution, with *L. nigrofrenata* being confined to Queensland. The broad head stripe and immaculate dorsal colour distinguish it from *L. pallida*.

Litoria pallida

Davies, Martin and Watson

Pale Frog

This small, ground-dwelling frog has unwebbed fingers with poorly developed discs, and the first finger is considerably longer than the second. There is a conspicuous dark stripe from the nostril to just in front of the eye. Behind the eye the stripe does not fully enclose the tympanum. There is also a distinct white or pale yellow stripe below the eye to the shoulder on the side of the head. The overall coloration ranges from a uniform grey or pale brown to a complex mottling of grey or brown.

SIZE: Males 27–34 mm; females 31–37 mm.

HABITAT: Ranges from dry, open forest to the periphery of the flood plains of northern Australia.

ADVERTISEMENT CALL: Males call from the edge of water between early November and early March. The call consists of an irregular and complex series of short and long components.

BREEDING AND LARVAE: From 50 to 350 eggs are laid in a surface layer. Tadpoles are mottled with brown and attain a maximum length of 49 mm. Development is completed in 8–10 weeks. The tooth formula is $\frac{2}{3}$.

SIMILAR SPECIES: The species is distinguished from *L. tornieri* by the broken (not continuous) dark stripe along the outer edge of the tibia, and the fact that it is never red. The only other similar species found in the same area is *L. inermis* which has a rough, irregularly warty skin and lacks the distinct, dark patch on the side of the head, whereas *L. nigrofrenata* is larger and has a continuous dark stripe behind the eye.

Litoria pallida, Fogg Dam, Northern Territory *(Photo by Gordon Grigg).*

Litoria personata, Cannon Hill, Northern Territory *(Photo by Bohdan Stankowich-Janusch).*

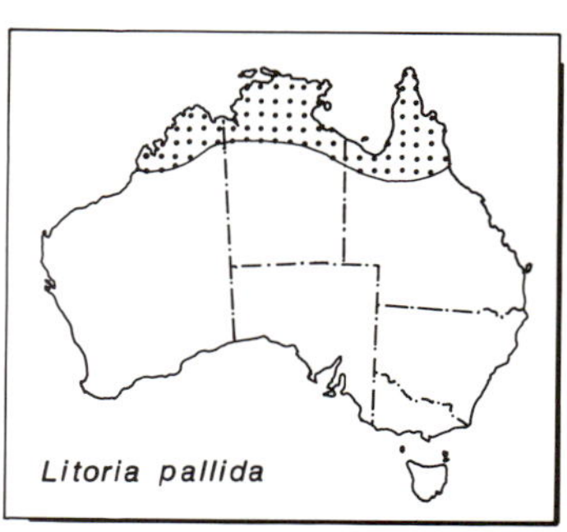

Litoria personata

Tyler, Davies and Martin

Masked Frog

This small, elongate frog has discs which are transversely oval and unwebbed, and toes approximately one-half webbed. The adult is dull brown or grey and bears a darker stripe on the side of the head, reaching the flanks. The stripe is poorly defined, not contrasting strongly with the overall coloration.

SIZE: Males 28–29 mm; females 31–33 mm.

HABITAT: Confined to the edge of escarpments of Arnhem Land, Northern Territory, where it seeks refuge beneath boulders and in cracks in the rock surface.

ADVERTISEMENT CALL: Males call from exposed positions upon the rock face in the early wet season (November to late January). Details of the call have not been reported.

BREEDING AND LARVAE: *Litoria personata* breeds in pools and rock crevices. The eggs are unknown, but the tadpole is the most distinctive of all Australian species, with brown upper portion and a golden stripe on each side. They attain a maximum length of 55 mm, and the tooth formula is $\frac{1\ ^{1}\ 1}{2\ _{1}\ 2}$.

SIMILAR SPECIES: The overall coloration and poorly defined lateral stripe create a resemblance to *L. watjulumensis* which is, however, a substantially larger species, commonly suffused with lemon on the undersurface and flanks whereas *L. personata* is colourless.

Litoria tornieri

(Nieden)

Tornier's Frog

At night this species can be a rich red, but by day dull brown or grey. The head stripe is dark and contrasts against the pale (daytime) background colour. The backs of the thighs are yellow, bearing a distinct longitudinal pattern of brown markings.

Litoria tornieri, light phase, Liverpool River, Northern Territory *(Photo by Gordon Grigg)*.

Litoria tornieri, dark phase, Melville Island, Northern Territory *(Photo by Margaret Davies)*.

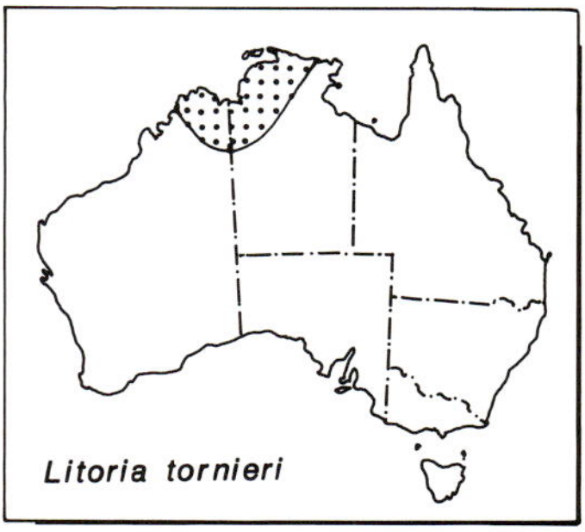

The discs on the fingers and toes are very small and no wider than the digits below them. The fingers are unwebbed and the toes are not fully webbed.

SIZE: Males 28–36 mm; females 27–34 mm.

HABITAT: Most common in open grasslands adjacent to temporarily flooded pools.

ADVERTISEMENT CALL: Males call from October to April from the ground beneath leaves or around the edges of shallow, temporarily filled pools. The call comprises a series of short and long notes, best described as a penetrating rattle.

BREEDING AND LARVAE: Eggs are laid in floating clumps in shallow temporary water. The maximum length is 44 mm. Development is completed in less than six weeks. The emerging tadpole is mottled with brown and has a tooth row formula of $\frac{2}{3}$.

SIMILAR SPECIES: This species can be distinguished from *L. inermis* by its smooth or finely tubercular skin, compared with the extremely warty skin of *L. inermis*. From *L. pallida* it differs in the rich red nocturnal coloration and the continuous stripe along the tibia.

Litoria watjulumensis

(Copland)

Watjulum Frog

This species was described at first as a sub-species of *L. latopalmata* but is in fact quite distinct. The dorsal colour is pale brown to dark grey, sometimes with some darker markings. The ventral surface is white or lemon except for the throat of breeding males, which is darkly mottled on the edge of the jaw. The dark head-stripe forms a narrow line from the snout to the eye and then widens to the full width of the tympanum, terminating at a point about midway between armpit and groin. The edge of the lower jaw is checked in dark brown and white. The legs are very long with small discs on the toes. Webbing between the toes extends to the disc of all toes except the fourth. The

Litoria watjulumensis, Mt Brockman, Northern Territory *(Photo by Harold Cogger)*.

posterior of the thigh is yellow with a heavy network of black markings. The first finger is slightly longer than the second.

SIZE: Males 33–38 mm; females 45–70 mm.

HABITAT: *L. watjulumensis* may be found near permanent and semi-permanent swamps, streams, and soaks and at the foot of rock outcrops in Western Australia and throughout the Northern Territory above the 500 mm isohyet.

ADVERTISEMENT CALL: A series of complex rattles and chuckles varying in sequence. There is no apparent pattern — just a highly variable number of different components.

BREEDING AND LARVAE: Between October and March males call close to water. Spawn is laid in floating clumps of 30–200 eggs. Development is completed in eight weeks. The tadpole is a mottled dark brown and the tooth formula is $\frac{1\ ^{1}\ 1}{3}$.

SIMILAR SPECIES: The broad band of dark coloration upon the sides of the head and flanks is shared amongst northern Australian hylids only with *L. personata*. However, *L. personata* is a much smaller species.

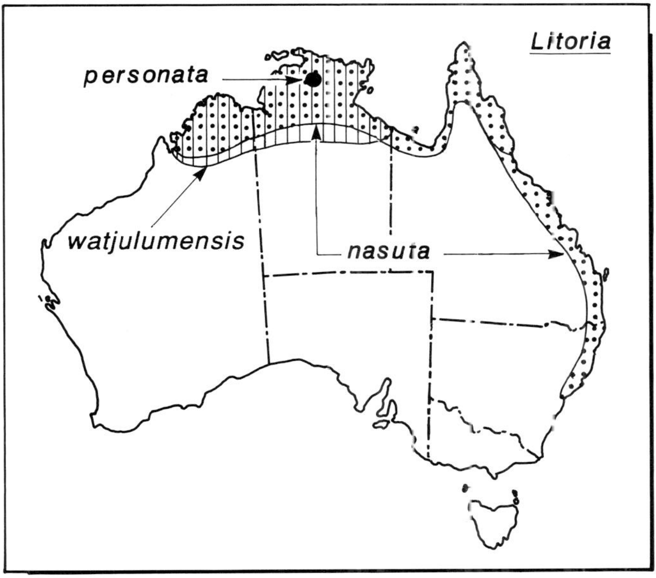
Litoria
personata
watjulumensis
nasuta

The *Litoria ewingii* Complex

This group of small to medium-sized tree frogs is a complex in the truest sense of the word and detailed analysis of mating calls, genetic compatibility experiments, and biochemical techniques have been applied to studies on these frogs. Murray Littlejohn and his colleagues have been working with these frogs for over 30 years and their work has contributed substantially to the resolution of taxonomic problems and evolutionary relationships between the species. Seven taxa are included in this complex, *L. ewingii, L. paraewingi, L. revelata, L. verreauxii verreauxii, L. verreauxii alpina, L. littlejohni* and *L. jervisiensis.*

Key to the *Litoria ewingii* Complex

1a. Axilla, groin and sides of thighs rich reddish-orange .... *L. littlejohni* (p. 86)
b. Axilla, groin and sides of thighs yellow or orange, with or without black markings 2

2a. Discs on fingers distinctly broader than adjacent finger width .... *L. revelata* (p. 89)
b. Discs on fingers only slightly broader than adjacent finger width 3

3a. Canthus rostralis straight; confined to central Victoria .... *L. paraewingi* (p. 88)
b. Canthus rostralis slightly curved; not confined to central Victoria 4

4a. Head broad, body squat 5
b. Head narrow, body slender 6

5a. Dorsal surfaces with brown patches upon pale green; skin warty; confined to Alps .... *L. verreauxii alpina* (p. 91)
b. Dorsal surfaces mottled with lighter and darker brown; skin smooth; not found in Alps .. *L. verreauxii verreauxii* (p. 92)

6a. Male call a series of 7–20 high notes lasting for 2–5 seconds .... *L. ewingii* (p. 83)
b. Male call 1–4 notes lasting for up to 2 seconds .... *L. jervisiensis* (p. 86)

Litoria ewingii

(Duméril and Bibron)

Ewing's Tree Frog, Whistling Tree Frog

This is a slender, medium-sized, brown tree frog. The head is broad with a rounded snout. A wide and undivided brown or grey band (sometimes indistinct) runs along the back from between the eyes to the cloaca. The remainder of the upper surface is pale brown except in south-east South Australia and western Victoria where the back can be green. There is a narrow black or dark brown stripe from the snout to the shoulder and a pale stripe on the upper jaw from beneath the eye to the angle of the jaw. The ventral surface is pale and granular and there is a distinct fold on the shoulder. The posterior surface of the thigh is yellow-orange, with or without circular black spots. Breeding males have a light grey-brown vocal sac. The toes are half webbed, with two joints and the disc of the fourth toe free. The fingers are free from web, and finger pads are slightly wider than the fingers. Breeding males have a brown nuptial pad on the thumb. The tympanum is distinct, and the vomerine teeth are between the choanae.

SIZE: Males 22–40 mm; females 32–46 mm.

HABITAT: *Litoria ewingii* occupies a wide variety of habitats from South Australia to southern New South Wales. It is usually found on the ground or in low vegetation on the banks of permanent streams or pools and commonly lives in suburban gardens.

ADVERTISEMENT CALL: A loud, harsh, high-pitched pulsing whistle, "weep-eep-eep", of about 10 to 20 short notes. Mating calls may be heard on even the coldest winter nights.

BREEDING AND LARVAE: This species may breed at any time of the year. The eggs are deposited in small clumps attached to submerged vegetation. The tadpoles are nearly transparent with only two bands of darker pigmentation along the body. The tail fin is very deep and terminates in a sharp point.

Litoria ewingii, Cann River, Victoria *(Photo by Gordon Grigg).*

Litoria ewingii, colour variant, Mortlake, Victoria *(Photo by John Wombey).*

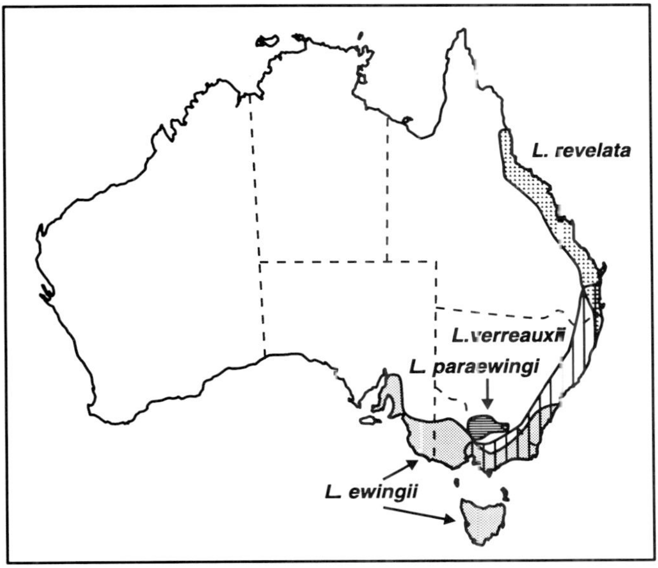

Total length reaches a maximum of 50 mm. Recently metamorphosed froglets are 11–14 mm in body length. Tadpole mouthparts $\frac{1\ ^{1}\ 1}{1\ _{2}\ 1}$.

SIMILAR SPECIES: *Litoria ewingii* is the only member of the complex in South Australia and western Victoria. In central Victoria it is replaced by the closely related *L. paraewingi*, from which it may be distinguished by male call and by the shape of the canthus rostralis (curved in *L. ewingii* and straight in *L. paraewingi*). In eastern Victoria and southern New South Wales confusion might arise with *L. verreauxii* which is distinguished most readily by its proportionately broader head and body. The latter has a more dumpy appearance whereas *L. ewingii* is distinctly slender.

Litoria jervisiensis

(Duméril and Bibron)

Jervis Bay Tree Frog

The overall colour is brown or dull grey. There is a broad, diffuse darker stripe commencing upon the eyelids and extending down the body, a narrower stripe extending from behind the eye to halfway along the body, and a white patch beneath the eye to beyond the angle of the jaw. There is a pale yellow patch in the axilla. The undersurfaces are creamish. Webbing between the fingers ranges from basal to one-third, and the toes are about three-quarters webbed.

SIZE: Males 29–37 mm; females 33–44 mm.

HABITAT: Found in coastal localities around permanent lagoons, hind-dune creeks, swamps and waterholes. The underlying substratum is nearly always loose sand.

ADVERTISEMENT CALL: Two to three high-pitched squeals over two seconds, repeated at short intervals. Males call from April to October.

BREEDING AND LARVAE: Confusion about the identity of *L. jervisiensis* (resolved by Arthur White and his colleagues), makes it uncertain whether the information about the breeding of this species involve it or the newly described *L. littlejohni*.

SIMILAR SPECIES: The species with which it has been confused (*L. littlejohni*) lacks a distinctive white patch beneath the eye, present in *L. jervisiensis*.

Litoria littlejohni

White, Whitford and Mahony

Littlejohn's Tree Frog

The head is broader than long with a rounded snout. There is a white patch beneath the eye. The dorsum is light brown with a broad dark undivided band (sometimes very indistinct) from between the eyes to the cloaca. The undersurface is paler: usually cream or yellowish. A black stripe runs from the nostril through

Litoria jervisiensis, Centennial Park, Sydney, New South Wales *(Photo by Arthur White).*

Litoria littlejohni, Darkes Forest, New South Wales *(Photo by Arthur White).*

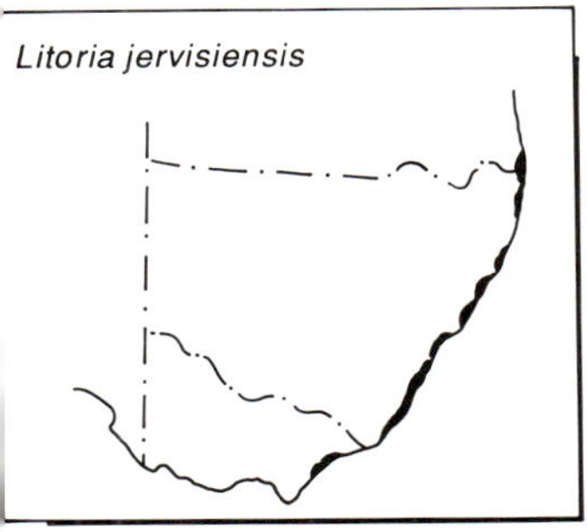

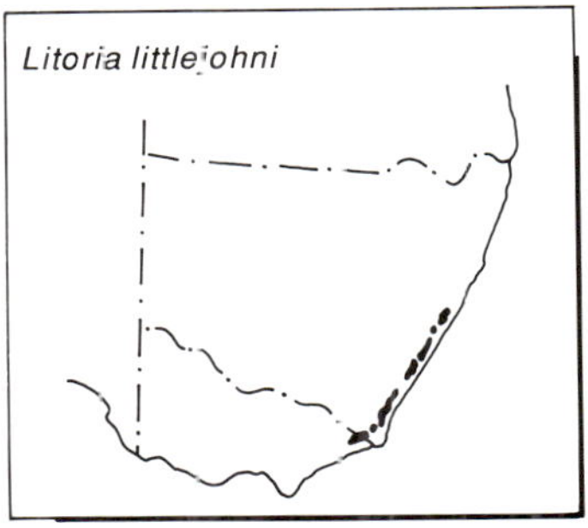

the eye to the shoulder. The sides and undersurface of the thigh, tibia, and the armpit and upper arm are bright reddish-orange. The dorsal skin texture tends to be warty and the undersurface granular. The toes are half-webbed and the fingers free. There are large discs on the fingers and toes.

SIZE: Males 38–56 mm; females 48–72 mm.

HABITAT: Wet or dry sclerophyll forest associated with sandstone outcrops, at altitudes of 280–1 000 m on the eastern slopes of the Great Dividing Range. *L. littlejohni* prefers flowing rocky streams and is often found with *L. citropa*. Specimens have also been collected in semi-permanent dams with some emergent vegetation.

ADVERTISEMENT CALL: A low reedy whistle, considerably lower in pitch and with longer note duration than others in the "*ewingii* complex". (Tape track 17).

BREEDING AND LARVAE: Large eggs are deposited in small clusters attached to submerged sticks and reeds. The tadpoles are almost black and grow to 50 mm. Larval life is probably about 12 weeks.

SIMILAR SPECIES: Distinguished from *L. jervisiensis* most readily by lack of a white patch beneath the eye.

Litoria paraewingi

Watson, Loftus-Hills, and Littlejohn

Victorian Frog

This is an example of what biologists call a "cryptic" species for it is so similar in appearance to *L. ewingii* that its recognition as a distinct species would have been unlikely without the aid of modern analytical techniques.

Litoria paraewingi is a little smaller than its sibling species. The dorsum varies from pale grey to brown with a dorsal band, broad and undivided, running from the eyes to the cloaca. There is a distinct white stripe under the eye. The posterior of the thigh is orange. Breeding males have a light to dark brown vocal sac and a brown nuptial pad on the first finger. The snout is pointed when seen from above due to the straight canthus rostralis.

SIZE: Males 21–28 mm; females 34–36 mm.

HABITAT: Similar to *L. ewingii.*

ADVERTISEMENT CALL: The call is very similar to the other species in this complex, but differs in having a longer note duration and fewer notes per call (three to eight).

BREEDING AND LARVAE: Similar to *L. ewingii.*

SIMILAR SPECIES: See *L. ewingii.*

Litoria revelata

Ingram, Corben and Hosmer

Revealed Frog

The overall colour ranges from a pale creamy brown to a reddish brown, commonly with a lyre-shaped darker marking or a mid-vertebral stripe on the back. There is a narrow dark brown stripe along the side of the snout to the eye, continuing behind the eye to the insertion of the arms. There is a narrow white stripe on the upper lip. The groin and posterior surfaces of the thighs are orange, bearing a few scattered, small black spots. The undersurfaces are pale cream. There is a trace of webbing between the fingers, and the toes are about one half webbed.

SIZE: Males 26–28 mm; females 30–36 mm.

HABITAT: From coastal swamps and ponds to montane forest, occurring in three discrete populations.

ADVERTISEMENT CALL: The male calls from reeds and grasses. The call is likened to a series of up to 19 high-pitched whirrings.

BREEDING AND LARVAE: Breeding takes place in ponds and swamps, but details of the spawn and development are unknown.

SIMILAR SPECIES: *Litoria revelata* resembles *L. ewingii* most closely in both size and colour pattern. It is distinguished by the breadth of the finger discs which are distinctly larger than the breadth of the adjacent finger in *L. revelata*, but scarcely broader in *L. ewingii.*

Litoria paraewingi, Narbethong, Victoria *(Photo by Murray Littlejohn).*

Litoria revelata, Warrie National Park, Queensland *(Photo by Bohdan Stankowich-Janusch).*

Litoria verreauxii alpina

(Fry)

Verreaux's Alpine Tree Frog

As the name suggests, this frog is found high in the Australian Alps. It is a striking species and there are several colour variations. It may reach a larger size and is more colourful than *L. v. verreauxii*. The dorsum is warty and the background colour may be green, brown, or grey. A brown or black dorsal band is present from the eyes to the cloaca but may be broken into a series of irregular blotches or stripes. The head is broad and short with a brown or black stripe from nostrils to shoulder and a light stripe on the upper jaw. The ventral surface is pale cream and granular. The toes are half-webbed and the fingers free. Discs on fingers and toes are small, as narrow as the digits. The tympanum is distinct, and the vomerine teeth are small and lie between the choanae.

SIZE: Males 31–33 mm; females 31–34 mm.

HABITAT: This is the predominant species in the Snowy Mountains. In winter they hibernate under stones and can tolerate snow cover.

ADVERTISEMENT CALL: Males call from the ground among low vegetation bordering streams and tarns. The call is a pulsing "wirr-wirr-wirr" similar to *L. v. verreauxii*, but usually deeper and with a slower pulse repetition rate, perhaps because of lower air temperatures. Great choruses of calling males may be heard in December and January. (Tape track 33).

BREEDING AND LARVAE: Similar to *L. v. verreauxii*.

SIMILAR SPECIES: See above.

Litoria verreauxii verreauxii

(Duméril)

Verreaux's Tree Frog

Litoria v. verreauxii is a medium-sized frog with a brown mid-dorsal band (which may be indistinct) extending from the eyes to the cloaca. The band may be divided for part or all of its length by a narrow lighter zone. The remainder of the dorsum is pale brown. A black stripe extends from the nostril through the eye to the shoulder. A white line runs along the hind margin of the upper jaw. The ventral surface is cream and granular. The groin is pale yellow with black spots, and the anterior and posterior of the thigh are orange. The discs on the fingers and toes are small. The toes are half-webbed, with two joints and the disc of the fourth toe free. The fingers are free from web. The tympanum is distinct.

SIZE: Males 27–36 mm; females 32–36 mm.

HABITAT: *Litoria v. verreauxii* occurs from central Victoria to south-east Queensland, but only in the vicinity of permanent water.

ADVERTISEMENT CALL: This is a rapid pulsing whistle "cree-cree-cree-cree", less harsh than that of *L. ewingii* and with a faster pulse rate. Males call from concealed sites on the banks of ponds. (Tape track 16.)

BREEDING AND LARVAE: Similar to *L. ewingii.*

SIMILAR SPECIES: *Litoria v. verreauxii* is likely to be confused with any of the smaller members of the *L. ewingii* complex, but will be distinguished if emphasis is placed on the overall body shape which is so much broader in this species.

The *Litoria peronii* Group

The *Litoria peronii* group includes six species of which three exist in Australia, one in Timor and the other two are confined to New Guinea. They all have extensive black and yellow markings in the axilla and behind the thighs. They are capable of a colour change from putty to dark grey or pale brown.

Litoria verreauxii alpina, Smiggins Holes, New South Wales *(Photo by Marion Anstis)*.

Litoria verreauxii verreauxii, Bald Mountain, Queensland *(Photo by Owen Kelly)*.

Key to the *Litoria peronii* Group

1a. Supratympanic fold edged with black; axilla black, or yellow with large black spots; backs of thighs bright yellow, heavily marked with black 2

b. Supratympanic fold not edged with black; axilla yellow with or without one or two black spots; backs of thighs yellow with fine black variegation .... *Litoria tyleri* (p. 98)

2a. Adult male greater than 42 mm S-V length; back warty; backs of thighs yellow or orange with extensive black variegation .... *Litoria peronii* (p. 94)

b. Adult males less than 42 mm S-V length; back smooth; backs of thighs black with a few yellow or orange spots .... *Litoria rothii* (p. 97)

Litoria peronii

(Tschudi)

Peron's Tree Frog

Body form identical to *L. tyleri* but most commonly grey rather than brownish and bearing numerous very small, pale green spots. The skin fold above the tympanum is edged with a fine black line. The backs of the thighs are heavily marked with black on yellow or orange. The population in Western Queensland has thigh coloration similar to *L. tyleri*, but its call is that of *L. peronii*.

SIZE: Males 44–53 mm; females 46–65 mm.

HABITAT: Found beneath loose bark on the trunk of large eucalypts flanking rivers and waterholes and, by night, on vegetation near the water. They call from vegetation around temporary ponds and permanent waterholes, often from well above the ground.

ADVERTISEMENT CALL: A long series of 29–50 explosive notes, like laughter, said by some to sound like a "maniacal cackle".

BREEDING AND LARVAE: Tadpoles are a pale golden yellow bearing dark lateral stripes. Maximum length recorded is 44 mm. Tadpole tooth formula is $\frac{1\;^{1}\;1}{1\;_{2}\;1}$.

Litoria peronii, Glen Lyon, Queensland *(Photo by Gordon Grigg).*

Litoria rothii, Mt Elliot National Park, Queensland *(Photo by Glenn Hoye).*

SIMILAR SPECIES: Over a portion of eastern Australia *Litoria peronii* coexists with and is likely to be confused with *L. tyleri* and *L. rothii*. The most readily observed distinguishing characteristic is the black line along the margin of the supratympanic fold in *L. peronii,* whereas in *L. tyleri* the fold is not pigmented. *L. rothii* is distinguished by its more extensive black pigmentation on the thighs and in the groin. Confusion with *L. rothii* is possible in the overlap zone between Rockhampton and Maryborough (see *L. rothii*).

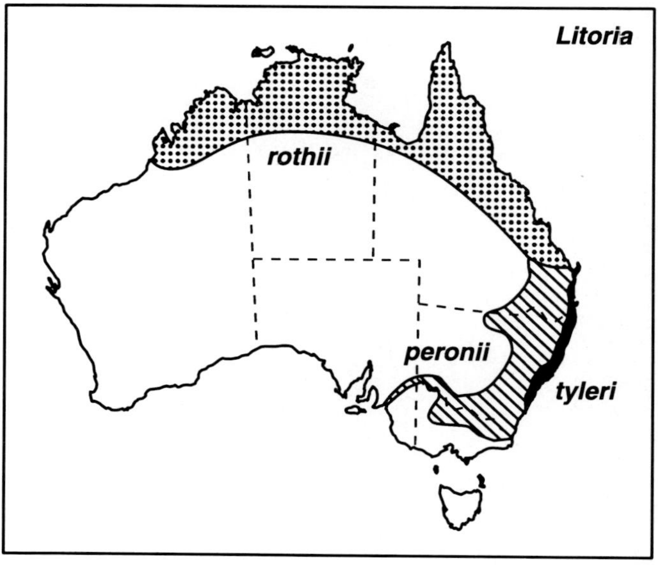

Litoria rothii

(de Vis)

Roth's Tree Frog

Body form identical to *L. tyleri*, but differing in coloration and habitat. The typical coloration is putty-like, without any markings, especially when basking. The same individual may be pale brown at night, heavily blotched with dark chocolate. The backs of the thighs are predominantly black, bearing a few, small yellow or orange patches. The upper half of the iris is deep wine-red.

SIZE: Males 37–48 mm; females 45–57 mm.

HABITAT: Vegetation adjacent to or overhanging rivers, streams or large temporary pools. Common on the leaves of *Pandanus*.

ADVERTISEMENT CALL: A series of 7–9 explosive, clearly separated notes resembling laughter.

BREEDING AND LARVAE: Breeding takes place from November to March. Spawn is deposited in small clumps in temporary pools. The tadpoles are pale yellow with dark brown stripes along the length of the body and a bright yellow spot anterior to the eyes. Tadpoles attain a maximum length of 58 mm and take a minimum of 65 days to complete development. The tadpole tooth formula is $\frac{1\ {}^{1}\ 1}{1\ {}_{1}\ 1}$.

SIMILAR SPECIES: Within northern Australia there are no other species resembling *L. rothii* with its distinctive eye and groin colour and there is only a small area of overlap with *L. tyleri* in south-east Queensland. *L. rothii* and *L. peronii* do overlap, however, between Rockhampton and Maryborough.

Litoria tyleri

(Martin, Watson, Gartside, Littlejohn and Loftus-Hills)
Tyler's Tree Frog

An elongate tree frog with a small head and large eyes, half-webbed fingers and two-thirds webbed toes. The tympanum is distinct. The back coloration ranges from mottled grey to brownish with or without numerous very small green spots. The skin fold above the tympanum is not edged with black. The axilla is yellow, with sometimes one or two small black spots; the back of the thighs and groin are yellow or orange and mottled with black.

SIZE: Males 43–48 mm; females 46–48 mm.

HABITAT: Most commonly encountered amongst vegetation adjacent to or overhanging permanent ponds.

ADVERTISEMENT CALL: A series of 5–10 explosive and separate notes resembling laughter, with fewer notes and less "explosive" than *L. peronii*. (Tape track 4.)

BREEDING AND LARVAE: *Litoria tyleri* breeds in ponds in the spring. Tadpoles are pale brown and the tooth formula is $\frac{1\ {}^{1}\ 1}{1\ {}_{2}\ 1}$.

SIMILAR SPECIES: Despite the overall similarity of all species in the group, *L. tyleri* can be distinguished most readily by examining the supratympanic fold, which lacks the black edge which is characteristic of *L. peronii* and *L. rothii*.

itoria tyleri, Lamington National Park, Queensland *(Photo by John Wombey).*

The *Litoria aurea* Complex

This group of large and extremely attractive species has one representative in northern Australia, two representatives in the south-west of Western Australia, and three in southeastern Australia. Although termed "tree frogs", they seldom venture into arboreal situations and have only very small adhesive pads on the fingers and toes. Frogs of the *L. aurea* complex are largely aquatic, and move overland only during prolonged rain periods. In this respect and in their overall body form they closely resemble the common European frogs of the genus *Rana*. The frogs are active during daylight hours and can be seen and heard in the water or on reeds soaking up warmth from the sun. Members of the complex are often called bell-frogs.

Some strange behavioural traits are found in members of this complex; when first picked up, they release copious amounts of acrid mucus and emit a scream of anguish which might be rather startling if one has not had previous experience of it. This is the "distress" call. Another of their habits, disconcerting to the naturalist trying to keep these frogs in captivity, is cannibalism.

The species in the south-east of the continent are among those species whose numbers have declined dramatically in the last ten years.

Key to the *Litoria aurea* Complex

1a. Toes fully webbed. Occurs in northern Australia .... *Litoria dahlii* (p. 104)
 b. Toes not fully webbed. Occurs in southeastern or southwestern Australia 2

2a. Confined to southwestern Australia 3
 b. Confined to southeastern Australia 4

3a. Pale mid-dorsal stripe. Posterior of thighs black with yellow spots .... *L. cyclorhyncha* (p. 103)
 b. Mid-dorsal stripe absent. Groin and posterior of thighs turquoise .... *L. moorei* (p. 106)

4a. Mid-dorsal stripe absent. Discs on toes wider than digits. Toes three-quarters webbed .... *L. aurea* (p. 100)
 b. Mid-dorsal stripe present. Discs smaller than, or equal to, width of toe. Toes with more than three-quarter webbing 5

5a. Discs on toes very small, less than width of toe. Toes fully webbed. Large yellow spots in groin and on posterior of thigh. Restricted to isolated populations in the New England Tableland and the Australian Capital Territory .... *L. flavipunctata* (p. 104)
 b. Discs on toes equal in width to digit. Toes a little less than fully webbed. Groin and thigh usually turquoise or with a few small pale yellow spots .... *L. raniformis* (p. 106)

Litoria aurea

(Lesson)

Green and Golden Bell Frog

Until recently *L. aurea* was a common species in coastal New South Wales but it has now declined dramatically in numbers, the reasons for which are unclear. Some would argue that the introduced fish *Gambusia affinis* is implicated. The dorsum is smooth, with a striking colour combination of green and gold. Usually the background colour is green with irregular large spots and stripes of gold. The amount of gold is variable and may even be absent. There is a white or yellow stripe on the side of the head, continuing along the body as a dorso-lateral fold. A thick black stripe usually underlines this pale band. The ventral surface is white and coarsely granular. The groin is a distinctive turquoise blue. The tympanum is quite distinct, and vomerine teeth are between the choanae. Fingers are free while the toes are about three-quarters webbed. The discs are fairly large.

itoria aurea, Smiths Lake, New South Wales *(Photo by Gordon Grigg)*.

.itoria aurea, a different colour pattern, Gosford. New South Wales *(Photo by Gordon Grigg)*.

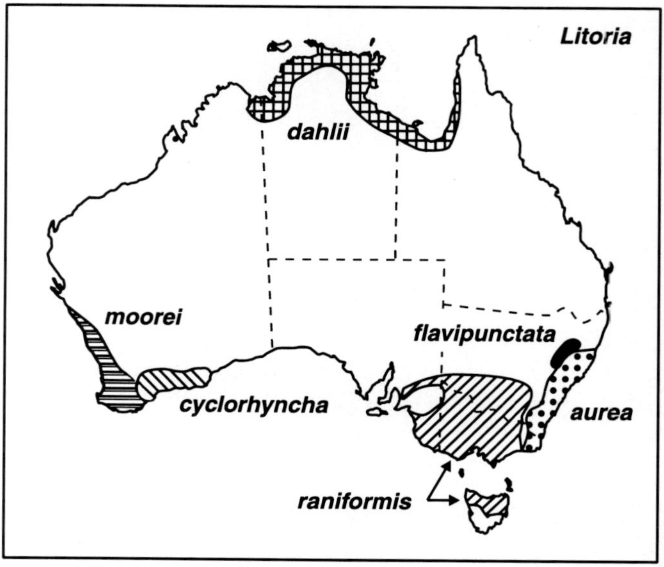

SIZE: Males 57–69 mm; females 65–108 mm.

HABITAT: The preferred habitat is large permanent ponds with an abundant growth of bulrushes near the bank. *Litoria aurea* often climbs among emergent vegetation at night while feeding.

ADVERTISEMENT CALL: A long, deep sonorous drone followed by several short grunts. (Tape track 12.)

BREEDING AND LARVAE: *Litoria aurea* may commence breeding about August. Males call while floating in the water and quite often the whole chorus is synchronized. Spawn is a loose mass of eggs among the floating vegetation. Larvae are typical of the genus: large size at metamorphosis, heavy pigmentation, and a fairly high tail fin.

SIMILAR SPECIES: In eastern populations, the features which distinguish *L. aurea* are the absence of a mid-dorsal stripe, moderately large discs, a smooth dorsum with green and iridescent gold coloration, and the absence of yellow spots on the flanks and groin.

Litoria cyclorhyncha

(Boulenger)

Spotted-thighed Frog

This is a distinctive member of the complex. The dorsum is green with irregular gold to brown blotches in varying amounts. There is no pale mid-dorsal stripe. The belly is light brown with dark marbling and is granular in texture. The dorso-lateral fold is usually absent. The most striking feature is the pattern in the armpits, groin, and thigh, which is black with large yellow spots. The discs are fairly small and the toes about three-quarters webbed. Fingers are free.

SIZE: Males 56–66 mm; females 63–77 mm.

HABITAT: Near swamps and other areas of permanent water.

ADVERTISEMENT CALL: A. R. Main described the call as "like the distant sound of wood being sawn". Identical to *L. moorei* (Roberts).

BREEDING AND LARVAE: In spring. Details yet to be described.

SIMILAR SPECIES: The bold pattern in the concealed areas of the groin serves to distinguish *L. cyclorhyncha* from any other species in Western Australia.

itoria cyclorhyncha, locality unrecorded, Western Australia *(Photo by Steve Donnellan)*.

Litoria dahlii

(Boulenger)

Dahl's Aquatic Frog

This is a moderately large aquatic species of northern Australia with slimy skin, long, pointed, unwebbed fingers and fully webbed toes. The body ranges from green or olive to dull slate, with irregular dark green spots or patches of golden-brown commonly bearing a green mid-vertebral stripe when in the slate phase. The tympanum is large and distinct, often partially covered by a fold.

SIZE: Males 49–63 mm; females 59–71 mm.

HABITAT: Closely associated with large expanses of water upon the flood plains of northern Australia. For much of the year the species can be found basking at the edge of the water.

ADVERTISEMENT CALL: A short bark.

BREEDING AND LARVAE: Tadpoles are dark slate and reach a maximum length of 57 mm. The tadpole tooth formula is $\frac{2}{3}$.

SIMILAR SPECIES: The size of this species, with complete webbing of the toes and lack of finger and toe discs, are a combination of features not shared by any other in northern Australia.

Litoria flavipunctata

Courtice and Grigg

Yellow-spotted Tree Frog

The general colour is almost identical to *L. raniformis* with green and gold marbling, black spots and a pale green mid-dorsal stripe. The distinctive features are the large yellow spots in the groin and on the posterior of the thigh. The toe discs are very small and are narrower than the width of the toe. The toes are fully webbed and the fingers are unwebbed.

SIZE: Males 58–73 mm; females 64–92 mm.

HABITAT: Found in similar situations to *L. raniformis*. Described from the New England Tableland, it is possible that the species also occurs in the Australian Capital Territory.

Litoria dahlii, Cannon Hill, Northern Territory (*Photo by Gordon Grigg*).

Litoria flavipunctata, West of Guyra, New South Wales (*Photo by Gordon Grigg*).

ADVERTISEMENT CALL: Very similar to *L. raniformis*. (Tape track 43.)

BREEDING AND LARVAE: Similar to *L. aurea*.

SIMILAR SPECIES: See *L. raniformis*.

Litoria moorei

(Copland)

Moore's Frog

This species, named after the American biologist John A. Moore, was previously known as *L. aurea* and is very similar in colour and markings. It is also closely related to *L. cyclorhyncha* and there is recent evidence that it hybridizes with that species. It ranges widely in colour from a uniform pale brown to green with darker brown or golden patches. A pale mid-dorsal stripe is usually present.

SIZE: Males 53–64 mm; females 65–74 mm.

HABITAT: Closely associated with permanent water, hiding beneath bark, logs or other cover.

ADVERTISEMENT CALL: Described as a long, low growl, of several parts, not unlike a motorcycle changing gears (Main). Call identical to *L. cyclorhyncha* (Roberts).

BREEDING AND LARVAE: The breeding biology and life history of this species are essentially the same as for the eastern forms, including the fact that *L. moorei* and *L. cyclorhyncha*, unlike most western frogs, are spring breeders.

SIMILAR SPECIES: *Litoria moorei* is likely to be confused only with *L. cyclorhyncha*. However, *L. moorei* has turquoise in the groin in contrast to *L. cyclorhyncha* which has large yellow spots on a black background.

Litoria raniformis

(Keferstein)

Southern Bell Frog

This is probably the largest species in the complex. The general colour is similar to *L. aurea*, but with a warty dorsum. There is always a pale green mid-dorsal stripe and often some large black spots in a roughly linear series on the back. The belly is coarsely granular and the groin and posterior of the thighs are turquoise.

Litoria moorei, Walpole, Western Australia *(Photo by Gordon Grigg) (Specimen courtesy Grahame Webb).*

Litoria raniformis, Hillston, New South Wales *(Photo by Glenn Hoye).*

Some specimens have a few small yellow spots on the posterior of the thigh. The pale dorso-lateral fold and the tympanum are quite conspicuous. Fingers are without webbing while the toes have a little less than a full web. The discs are equal in width to the digits.

SIZE: Males 55–65 mm; females 60–104 mm.

HABITAT: Similar situations to *L. aurea* but not climbing in emergent vegetation because of its lesser agility.

ADVERTISEMENT CALL: Very similar to *L. aurea* except that the short grunts usually preface the long drone. (Tape track 42).

BREEDING AND LARVAE: As for *L. aurea*. Tadpoles can attain a length of 100 mm. The tadpole tooth formula is $\frac{1\ {}^{1}\ 1}{1\ {}_{2}\ 1}$.

SIMILAR SPECIES: This species is very closely related to *L. flavipunctata* but can be distinguished by having slightly larger discs and lacking extensive yellow spots on the thighs and in the groin.

The Green Tree Frogs

There are 15 species of predominantly green hylid frogs that have large discs on the tips of their fingers and toes and which are found commonly in trees, shrubs or elsewhere above the ground. Without implying that they are necessarily closely related, it is convenient to group them for the purposes of identification, as follows:

1. *Litoria caerulea* group: *L. caerulea, L. gilleni, L. splendida, L. cavernicola* and *L. infrafrenata*.
2. *Litoria bicolor* group: *L. bicolor, L. fallax, L. cooloolensis* and *L. olongburensis*.
3. *Litoria citropa* group: *L. citropa, L. subglandulosa, L. phyllochroa* and *L. pearsoniana*.
4. *Litoria chloris* group: *L. chloris, L. xanthomera* and *L. gracilenta*.

Apart from *L. caerulea* and *L. gilleni* , all of the green tree frogs are restricted to the coast and mountains of northern and eastern Australia, whilst *L. bicolor, L. caerulea, L. gracilenta,* and *L. infrafrenata* occur also in New Guinea. It should be remembered that almost all frogs are capable of some colour change and a normally bright green frog may become olive or dark brown, especially in daylight hours.

Key to groups of Green Tree Frogs

1a. Dark stripe from nostril to eye, expanding behind the eye to the flanks .... *Litoria citropa* group (p. 124)
b. No dark stripe from nostril to eye and expanding behind the eye 2

2a. Body length less than 35 mm . *Litoria bicolor* group (p. 117)
b. Body length greater than 35 mm 3

3a. Prominent glands on the head, or a distinctive white stripe on the upper lip .... *Litoria caerulea* group (p. 109)
b. No prominent glands on the head and no distinctive white stripe on the upper lip .... *Litoria chloris* group (p. 130).

Litoria caerulea Group

Key

1a. White stripe along lower jaw to shoulder .... *Litoria infrafrenata* (p. 114)
b. Lacking white stripe from lower jaw to shoulder 2

2a. Possessing parotoid glands or glands upon the entire top of the head 3
b. Lacking parotoid or other glands . *Litoria cavernicola* (p. 112)

3a. Vast gland covering entire head; found only in Kimberley Division of Western Australia and the adjacent Northern Territory .... *Litoria splendida* (p. 116)
b. Possessing parotoid glands only 4

4a. Confined to the vicinity of Alice Springs; males less than 65 mm .... *Litoria gilleni* (p. 112)
b. Extensively distributed in northern Australia but not found near Alice Springs; males greater than 65 mm ... . *Litoria caerulea* (p. 110)

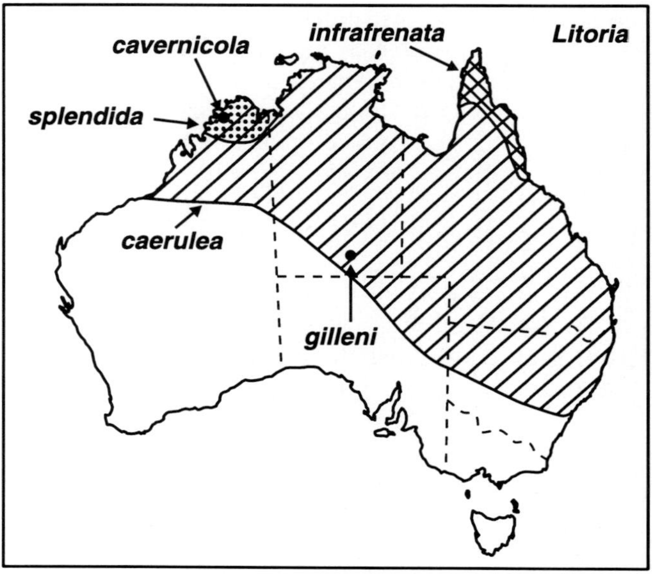

Litoria caerulea

(White)

Green Tree Frog

This attractive green frog is one of the most commonly encountered frogs in Australia because it inhabits letterboxes, toilet bowls and cisterns, meter boxes, and bathrooms. The dorsal surface is smooth with large glands on the back of the head and many tiny pores in the skin. Fingers and toes have large discs and well developed webbing. The colour is usually bright green but may change to olive-brown. Many specimens have small distinct white spots on the back, sides, and limbs. The ventral surface is white and granular. A large fleshy fold overhangs the distinct tympanum and the vomerine teeth are between the choanae. The back of the thighs varies from yellow to maroon. The iris is pale gold. *Litoria caerulea* is the most popular species for keeping in captivity, but it is now protected in all States except South Australia.

SIZE: Males 66–77 mm; females 60–110 mm.

HABITAT: Apart from its domestic environment, *L. caerulea* is found on rocks and trees and on the ground near streams and swamps.

ADVERTISEMENT CALL: A deep "crawk" repeated continuously. (Tape track 14).

BREEDING AND LARVAE: Clumps of 200–2 000 eggs are deposited in still water from November to February. The tadpoles are mottled brown or green and reach a maximum length of 45–100 mm. Development is completed in about six weeks. The tadpole tooth formula is $\frac{2\ 2}{3}$.

SIMILAR SPECIES: Two species share with *L. caerulea* the enlarged head glands: *L. gilleni* and *L. splendida*. *Litoria gilleni* tends to be smaller and replaces *L. caerulea* around Alice Springs, *L. splendida* has skin glands covering the entire head and not just the posterior portion. *L. cavernicola* resembles *L. caerulea* but lacks the fold on the side of the head overhanging the tympanum and has vomerine teeth behind, not between, the choanae.

Litoria caerulea, Ourimbah, New South Wales (*Photo by Gordon Grigg*).

Litoria cavernicola

Tyler and Davies

Cave-dwelling Frog

This poorly known species was first described in 1979 and resembles a small and more slightly built *L. caerulea*. The overall coloration is dull green or a brownish green. The skin is distinctly granular. There are small discs on the tips of the fingers and toes and slight webbing at the base of the fingers, and extensive webbing between the toes. The tympanum is large and distinct but the upper rim is commonly hidden beneath a fold of skin. Vomerine teeth are behind the choanae.

SIZE: Males 44–51 mm; females 50–57 mm.

HABITAT: Confined to caves in sandstone gorges and refuges between boulders in rock plateaus on the Mitchell Plateau of the Kimberley Division of Western Australia.

ADVERTISEMENT CALL: Unknown

BREEDING AND LARVAE: Unknown.

SIMILAR SPECIES: The only green tree frogs in the area where *L. cavernicola* is found are the tiny and slender *L. bicolor*, *L. caerulea* and possibly *L. splendida*. *L. cavernicola* is substantially larger than *L. bicolor*. From a juvenile *L. splendida* the absence of the vast head glands and the cream spots would distinguish *L. cavernicola*. Confusion might arise with a small *L. caerulea*, and the simplest means of distinguishing the two would be to note that *L. cavernicola* has a proportionately larger tympanum, a granular skin (smooth in *L. caerulea*), a larger head and vomerine teeth behind the choanae.

Litoria gilleni

(Spencer)

Centralian Tree Frog

There has been dispute about whether *Litoria gilleni* is specifically different from *L. caerulea*. Biochemical evidence obtained recently is quite convincing that *L. gilleni* should be considered a distinct species, confined to the vicinity of the MacDonnell Ranges in

Litoria cavernicola, Mitchell Plateau, Western Australia *(Photo by Mike Mahony)*.

Litoria gilleni, Alice Springs, Northern Territory *(Photo by Glenn Hoye)*.

the Northern Territory. Externally *L. gilleni* resembles *L. caerulea* closely but tends to have more numerous white spots upon the dominant green or olive background.

SIZE: Males 40–62 mm; females 55–80 mm.

HABITAT: Confined to rock gorges where there are permanent waterholes or even temporary water. The frogs hide in the cool spaces beneath boulders during the day.

ADVERTISEMENT CALL: A harsh, slowly repeated barking call resembling the call of *L. caerulea*.

BREEDING AND LARVAE: Nothing appears to have been published about the breeding biology of this species.

SIMILAR SPECIES: Despite the close similarity to *L. caerulea*, identification is a simple matter because *L. gilleni* is restricted to central Australia, and the only other tree frog in the area is *L. rubella*, which is small and brown and could not be confused with it.

Litoria infrafrenata

(Günther)

White-lipped Tree Frog

This large and spectacular frog grows to 135 mm. Its dorsal surface is uniform green or bronze, the ventral surface white or off-white. The throat of the males is greenish. There is a conspicuous white stripe around the lower jaw continuing back to the shoulder and white stripes (salmon pink in breeding males) on the trailing edges of the tibia (lower leg). The fingers and toes have large discs. The fingers are half-webbed and the toes have an almost complete web. The tympanum is large and distinct, the central portion usually having some green pigmentation. A fold develops above the tympanum in large specimens.

SIZE: Males 62–102 mm; females 73–135 mm.

HABITAT: *L. infrafrenata* can be found in most non-rainforest habitats on the coast of northeastern Queensland, including in suburban areas. It is commonly transported accidentally to southern States in boxes of bananas.

itoria infrafrenata, Cairns, Queensland *(Photo by Gordon Grigg)*.

itoria infrafrenata, early metamorph just left the water, Cooktown, northern Queensland *(Photo by Gordon Grigg)*.

ADVERTISEMENT CALL: A loud, harsh double note resembling the barking of a large dog.

BREEDING AND LARVAE: Detailed accounts of the development of this species, following several spawnings in the Royal Melbourne Zoo, have been reported by Chris Banks and his colleagues. Up to 4 100 eggs are laid in dumbbell-shaped clumps. The tadpoles are dark brown with a single cream stripe on each side of the head, body and tail. Development is completed in about eight weeks. The tadpoles attain a maximum length of 54 mm and the tooth formula is $\frac{1\ ^{1}\ 1}{3}$ ·

SIMILAR SPECIES: Confined in Australia to Cape York Peninsula and the wet tropics of north-east Queensland, *L. infrafrenata* is likely to be confused only with *L. caerulea*. It is readily distinguished by the white stripe on its lower jaw.

Litoria splendida

Tyler, Davies and Martin

Magnificent Tree Frog

This is a large and handsome species characterized by having a large gland upon the entire dorsal surface of the head. The head and body are a rich green and bear small, scattered sulphur-coloured patches. The backs of the thighs are yellow or orange. The fingers and toes have large discs and the toes are fully webbed. The tympanum is large and distinct except for the upper rim which is hidden beneath the outer edge of the head gland.

SIZE: Males 88–104 mm; females 94–106 mm.

HABITAT: During the day the species enters caves, crevices beneath boulders, and can be encountered around buildings, public toilets, etc.

ADVERTISEMENT CALL: The call resembles that of *L. caerulea*: a series of deep barks.

BREEDING and LARVAE: Details are unknown. It is likely that *L. splendida* breeds early in the wet season — December/January.

Litoria splendida, Lake Argyle, Western Australia (*Photo by Margaret Davies*).

SIMILAR SPECIES: Because *L. splendida* is confined to the north-west of Western Australia and the adjacent margin of the Northern Territory, the closely related *L. caerulea* is the only species with which *L. splendida* could be confused. Both species are large and green and, in that area, individuals of each have yellow thighs. The main distinguishing features are the huge gland upon the top of the head of *L. splendida* (glands in *L. caerulea* are confined to the area behind the head), and the sulphur-coloured patches on the back of *L. splendida*.

The *Litoria bicolor* Group

Key

The four species *Litoria bicolor, L. fallax, L. cooloolensis* and *L. olongburensis* are all small (to 30 mm) green tree frogs. They are usually found close to water, often among the emergent vegetation.

1a. Overhanging, pointed snout. Metallic peacock blue on thigh and groin. Thickish lateral white line from eye to groin, broken and globular posteriorly .... *L. olongburensis* (p. 122)
b. Not as above 2

2a. Dorsum colour very variable, from bright green to dark green to brown to bronze. Often with dark brown spots and reticulations which may be very fine. Purplish-brown streak bordering the green colour of the upper thigh .... *L. cooloolensis* (p. 120)
b. Not as above 3

3a. Body length to head width ratio greater than 3.5. Thin bronze stripe from tip of snout, through nostril and eye, to groin .... *L. bicolor* (p. 118)

b. Body length to head width ratio less than 3.5. Generally larger than the others. No stripe from head to groin .... *L. fallax* (p. 120)

Litoria bicolor

(Gray)

Northern Dwarf Tree Frog, Green Reed Frog

This species and *L. fallax* are closely related and are morphologically similar. They were not given separate specific status until 1969.

Litoria bicolor is a small slender frog with a narrow head. The dorsal surface is green, bounded by a thin bronze stripe on each side from the shoulder to the groin, or it may be entirely pale bronze. The iris is gold. The ventral surface is white. A white stripe begins beneath the eye and continues past the shoulder and along the sides. There is a flash of orange in the groin and thigh. The fingers are slender and are slightly webbed, and the toes heavily webbed.

SIZE: Males 23–27 mm; females 25–29 mm.

HABITAT: *Litoria bicolor* can be found around permanent or semi-permanent streams and marshy areas from northern Western Australia to Bowen, northern Queensland. In the area of overlap with *L. fallax* it is found at lower elevations.

ADVERTISEMENT CALL: The calls of *L. bicolor, L. fallax* and *L. cooloolensis* are very similar and consist of a very complex high-pitched "wree-e-eck pippip". The differences are in the length of the first phase and the interval between phases. The first phase of the call of *L. bicolor* is twice as long as that of *L. fallax,* and the second phase begins before the first has ended, giving a rattling tail to the call. The call of *L. cooloolensis* is similar to that of *L. fallax*.

BREEDING AND LARVAE: Calling begins with the summer rain (December) and small clumps of pigmented eggs are laid attached to submerged vegetation. The tadpoles are very distinctive, having a high, crested tail and extensive pigmentation on both the muscular portion and the fin of the tail. They may reach 55 mm.

Litoria bicolor, Townsville, Queensland *(Photo by Steve Richards)*.

SIMILAR SPECIES: The morphological features which separate *L. bicolor* and *L. fallax* are the body length to head width ratio (*L. bicolor*, >3.5; *L. fallax*, <3.5), and the thin bronze-coloured lateral stripe of *L. bicolor*. *L. cooloolensis* has a mottled dorsum and a dark stripe on the posterior of the thigh. *L. olongburensis* has metallic blue coloration on the thigh and groin. *L. microbelos*, another species with which confusion is possible, is slightly smaller, has a more pointed snout, and less webbing between the toes.

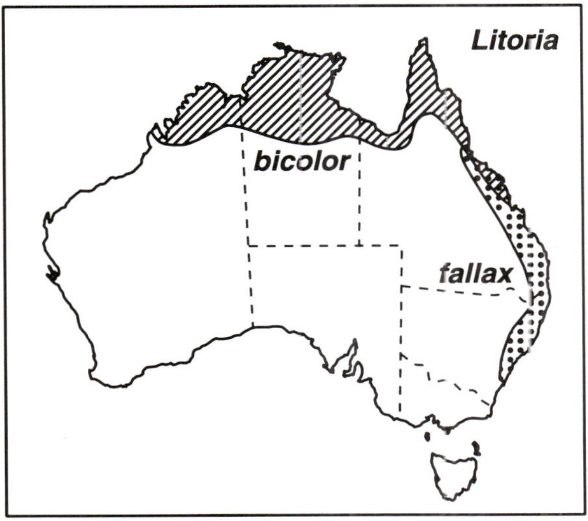

Litoria cooloolensis

Liem

Cooloola Tree Frog, Cooloola Sedge Frog

This attractive little frog remained unnoticed until 1974 when it was found at Cooloola and on Fraser Island, southeastern Queensland. The dorsum, unlike that in other species of the complex, is mottled and reticulated with fine to larger brown splotches on a green background. The ventral surface is cream. The side of the head is marked with a thin dark stripe from the nostril to the eye and a cream stripe from below the eye to the groin. The posterior surface of the thigh is orange with a thin purplish-brown stripe bordering the green dorsal colour of the thigh. Fingers are about one-third webbed and the toes are fully webbed.

SIZE: Males 21–26 mm; females 21–27 mm.

HABITAT: *L. cooloolensis* is found in the sandy coastal and island lakes and wallum creeks of southeastern Queensland and co-exists with *L. fallax* and *L. olongburensis*. It appears to prefer dense reed beds.

ADVERTISEMENT CALL: Very similar to *L. fallax*.

BREEDING AND LARVAE: Males call from August to May. Spawn is very similar to *L. bicolor* and *L. fallax*. Larvae are similar in size and shape although not as brightly coloured as in the other two species.

SIMILAR SPECIES: See *L. bicolor*.

Litoria fallax

(Peters)

Eastern Dwarf Tree Frog

Litoria fallax is one of the smallest of the green tree frogs. It is of slender build. Like others in the *L. bicolor* complex, it may be uniform green, uniform brown, or green with a pale brown or golden back. The ventral surface is white and granular. There is always a white stripe from the angle of the jaw, below the eye, to the shoulder. The iris is golden and the anterior and posterior surfaces of the thigh are orange. Webbing between the toes is well developed but the fingers have only a trace of webbing.

itoria cooloolensis, Fraser Island, Queensland *(Photo by Gordon Grigg)*
(Specimen courtesy Christina James)

itoria fallax, Brisbane, Queensland *(Photo by Lyn Beard)*.

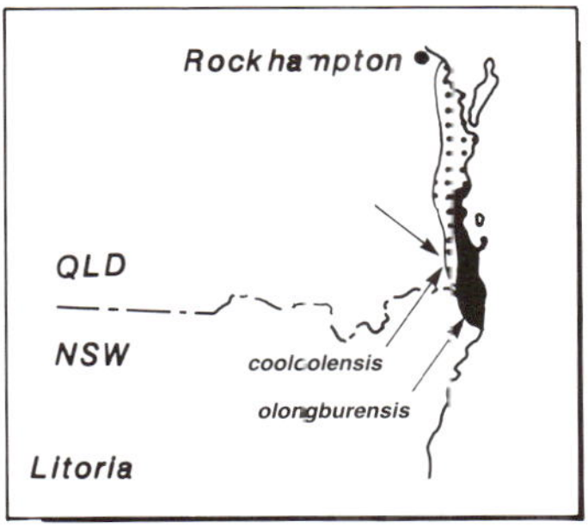

HABITAT: *Litoria fallax* is a coastal species, ranging from northern Queensland to central New South Wales. It is usually found in the vegetation bordering swamps, streams, ponds and farm dams, often in large numbers.

SIZE: Males 22–26 mm; females 25–32 mm.

ADVERTISEMENT CALL: A very high-pitched, upwards inflecting "wreek-pip-pip". The first component is drawn out, the latter part short and staccato. See *L. bicolor*. They call from October to April from emergent vegetation. (Tape track 10.)

BREEDING AND LARVAE: See *L. bicolor*. The tadpole tooth row formula is $\frac{1\ {}^{1}\ 1}{3}$.

SIMILAR SPECIES: See *L. bicolor*.

Litoria olongburensis

Liem and Ingram

Olongburra Frog, Wallum Sedge Frog

A small and slender tree frog with a narrow head and projecting snout. The fingers and toes are narrow, with the webbing basal between the fingers, and reaching midway up the penultimate portion of the fifth toe. The dorsal surface is dull green to brown/bronze, with a dark brown stripe along the side of the head and body and between the eye and the nostril. Beneath this dark stripe a narrow cream/white stripe passes from beneath the eye to the flanks, where it breaks up into thick, globular spots. The undersurface is white and granular, with the throat stippled in brown.

SIZE: Males approx. 25 mm; females 27–31 mm.

HABITAT: Found around a wide variety of water bodies where there is adequate fringing vegetation to provide cover: sand dune lakes and wallum creeks, mostly in areas of sandy soils where the water tends to be acidic.

ADVERTISEMENT CALL: A long, quiet buzz.

BREEDING AND LARVAE: Details have yet to be published.

SIMILAR SPECIES: *L. olongburensis* has a clear resemblance to *L. fallax* and *L. cooloolensis* from the same area. It is distinguished most readily by blue in the groin and the back of the thigh, and the overhanging snout.

.itoria olongburensis, Cooloola, Queensland *(Photo by Gordon Grigg).*

.itoria olongburensis, showing thigh coloration *(Photo by Gordon Grigg) (Specimen courtesy of Christina James).*

The *Litoria citropa* Group

Key

1a. Large gland beneath chin 2
 b. No large gland beneath chin 3

2a. Confined to New England Tablelands; males 34–40 mm; females 46–50 mm; tadpole lacks teeth .... *Litoria subglandulosa* (p. 129)
 b. Extends from southern New South Wales to Victoria; males 47–57 mm; females 56–57 mm; tadpole possesses teeth .... *Litoria citropa* (p. 124)

3a. Tympanum green .... *Litoria phyllochroa* (p. 128)
 b. Tympanum brown .... *Litoria pearsoniana* (p. 126)

Litoria citropa

(Tschudi)

Blue Mountains Tree Frog

This species has been described as the most beautiful tree frog in Australia and many specimens certainly live up to this reputation. The top of the head and back are light to dark brown, often mottled in black. There is nearly always a flash of green on the sides of the head and on the flanks. Ventral surfaces are white and granular. A black line runs from the nostrils, through the eye and over the tympanum (which is usually brown) to the groin. There is usually a thin white stripe above the dark stripe. These two bands break into a beautiful mottled pattern on the sides. The flash colours are bright orange-red. Fingers and toes have well developed discs. The fingers are free from webbing and the toes are half-webbed.

SIZE: Males 47–57 mm; females 56–57 mm.

HABITAT: This species is rarely seen. The preferred habitat appears to be rocky flowing streams in heavily forested areas.

ADVERTISEMENT CALL: A sharp harsh scream followed by a soft trill.

Litoria citropa, Helensberg, New South Wales *(Photo by Gordon Grigg)*.

BREEDING AND LARVAE: Calling males and clasping pairs have been seen in September and October. Males call from a variety of positions, most commonly from rocks and vegetation within a few metres of the water. Eggs are deposited on rocks in fast-flowing streams. The tadpole tooth formula is $\frac{1\ {}^{1}\ 1}{1\ {}_{2}\ 1}$.

SIMILAR SPECIES: *Litoria subglandulosa* appears to be the closest relative. *Litoria citropa* can be distinguished by its brown dorsum and flash colours, a brown tympanum (green in *L. subglandulosa*) and less distinct stripe on the upper jaw. The tadpoles are quite dissimilar as those of *L. subglandulosa* lack tooth rows.

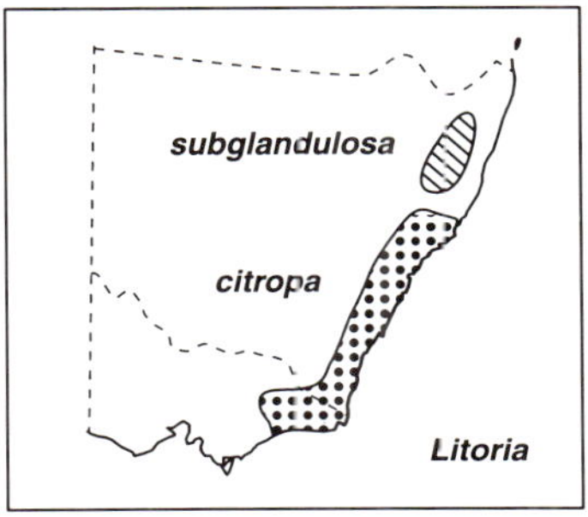

Litoria pearsoniana

(Copland)

Pearson's Green Tree Frog

This species is very close to *L. phyllochroa* and can be distinguished by mating call and distribution. Dorsal colour is uniform leaf green, and the ventral colour is white. A thin pale yellow or white stripe underlined in black begins at the nostril, passes through the eye, and continues to the shoulder. The tympanum is brown. A fairly broad dark brown or black band commences just behind the eye and continues back, widening to take in most of the side. The groin and posterior of the thigh are brick-red. Fingers have only a trace of webbing and toes are fully webbed.

SIZE: Males 24–29 mm; females 30–37 mm.

HABITAT: *L. pearsoniana* seems to prefer shaded rainforest gullies in association with fast flowing rocky streams. By day, and during winter, specimens can be found under stones at the water's edge. During the breeding season adults are active on the ground and on low shrubs bordering streams.

ADVERTISEMENT CALL: Adult males call from rocks or low vegetation from September to February. It is a beautiful chuckling sound and described as "week-kuk-kuk", taking from three-quarters to almost two seconds.

BREEDING AND LARVAE: Oviposition occurs in shallow water in spring and summer in the quieter pools. Egg-masses of 360–730 eggs are cemented to rocks, twigs or the pool floor. It attains a maximum length of 30 mm. The tadpoles metamorphose in 2–2½ months. The tadpole is dark brown and the tooth formula is $\frac{1\ ^{1}\ 1}{1\ _{2}\ 1}$.

SIMILAR SPECIES: *Litoria pearsoniana* closely resembles *L. phyllochroa*. They occur together between Taree and Coffs Harbour. The male calls are distinctive. Also *L. pearsoniana* lacks a fine gold and black line running from the nostril to the eye and on to the supratympanic fold.

Litoria pearsoniana, Cunningham's Gap, Queensland *(Photo by Gordon Grigg).*

Litoria phyllochroa, Belrose, New South Wales *(Photo by Gordon Grigg).*

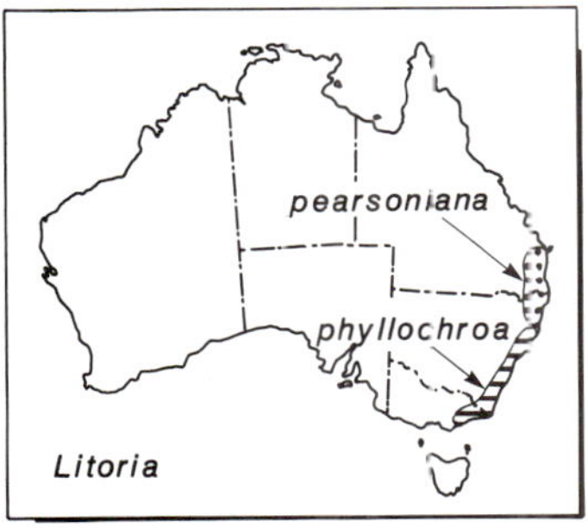

Litoria phyllochroa

(Günther)

Leaf Green Tree Frog

Litoria phyllochroa is normally very bright green but can change rapidly to a dull olive-brown. The ventral surface is white and granular. Posterior of thigh, armpit, and groin are brick-red. A conspicuous white stripe, underlined in black, runs from the nostril through the eye to the shoulder. A variety from the northern tablelands of New South Wales is smaller, may lack the flash colours, and often has a number of small black dots scattered over the back. There is a third form from Victoria in which the head-stripe continues past the shoulder and forms a broad mottled band along the sides to the groin similar to *L. citropa*. This form has a distinctive advertisement call and probably warrants separate specific status.

SIZE: Males 27–32 mm; females 30–41 mm.

HABITAT: *L. phyllochroa* occurs along the east coast from northern New South Wales to the Gippsland (Victorian) coast and as far west as the Great Dividing Range. It is usually found on low shrubs and vegetation beside creeks and ponds.

ADVERTISEMENT CALL: This has been described as a whirring "wrk r-r-r-rk w-r-r-r-rk", often heard during the day. (Tape track 26).

BREEDING AND LARVAE: Spawning occurs from September to April. Clumps of small eggs are attached to submerged vegetation. Larvae hatch in three to four days. Tadpoles are of the typical swimming type, reaching a length of 38 mm and take about twelve weeks until metamorphosis.

SIMILAR SPECIES: Apart from the distinct Victorian form described above, the most closely similar species is *L. pearsoniana*. However, *L. phyllochroa* can be distinguished by the absence of a dark lateral stripe and a green tympanum. They may also be separated by mating call and locality data.

Litoria subglandulosa, Bulga State Forest, New South Wales *(Photo by Marion Anstis).*

Litoria subglandulosa

Tyler and Anstis

Glandular Frog

For many years, this species was considered to be a form of *L. citropa*. It is now clear that it is a distinct species with some unique features, particularly in its larval form. The dorsum is either green or olive-brown. The sides are nearly always green. The head-stripe is very similar to *L. citropa* although the white stripe on the upper jaw is very much reduced or absent in *L. citropa*. The posterior surface of the thigh is reddish-brown. Fingers are free from webbing, toes are nearly fully webbed.

SIZE: Males 34–40 mm; females 46–50 mm.

HABITAT: *Litoria subglandulosa* lives near small streams in northern New South Wales, often in disturbed grazing lands and extends just into Queensland.

ADVERTISEMENT CALL: A series of several, moderately low-pitched notes initially increasing in volume and rate, and finally slowing, "orak-orak-orak . . .".

BREEDING AND LARVAE: This frog breeds in the spring (October–November). The larvae are unlike any other species of *Litoria* in that teeth and the horny beak are absent and there is a series of long papillae or tentacles around the mouth. Tadpoles reach a maximum length of 34 mm and take two to four months to complete their development.

SIMILAR SPECIES: See *L. citropa*.

Litoria chloris Group

Key to *Litoria chloris* Group

1a. Pale yellow or white stripe from nostril to eye .... *Litoria gracilenta* (p. 132)
 b. No pale stripe from nostril to eye 2

2a. Back of thighs purple .... *Litoria chloris* (p. 130)
 b. Back of thighs orange .... *Litoria xanthomera* (p. 134)

Litoria chloris

(Boulenger)

Red-eyed Green Tree Frog

Litoria chloris is a strikingly coloured, moderately large frog (body length to 65 mm) which spends much of its life high in the trees and is therefore seen infrequently, usually in association with heavy rain. The dorsal surface is green and shows little variation. The ventral surface is a vivid deep yellow with a coarse granular texture. The only green pigmentation occurring on the limbs is on the upper surface of the forearm and the tibia. There is usually a thin strip of green extending to the upper arm and down to the hand and a similar narrow band on the upper surface of the thigh. The rest of the limb is yellow. The posterior of the thigh is purple, and the iris is bright orange. Hands and feet have very large discs and well-developed webs. The tympanum is distinct and smooth.

SIZE: Males 54–62 mm; females 58–68 mm.

HABITAT: *Litoria chloris* is usually associated with coastal rainforest or wet sclerophyll forest and, consequently, has a scattered distribution along the Queensland–New South Wales coast extending as far north as

itoria chloris, Ourimbah, New South Wales *(Photo by Gordon Grigg)*.

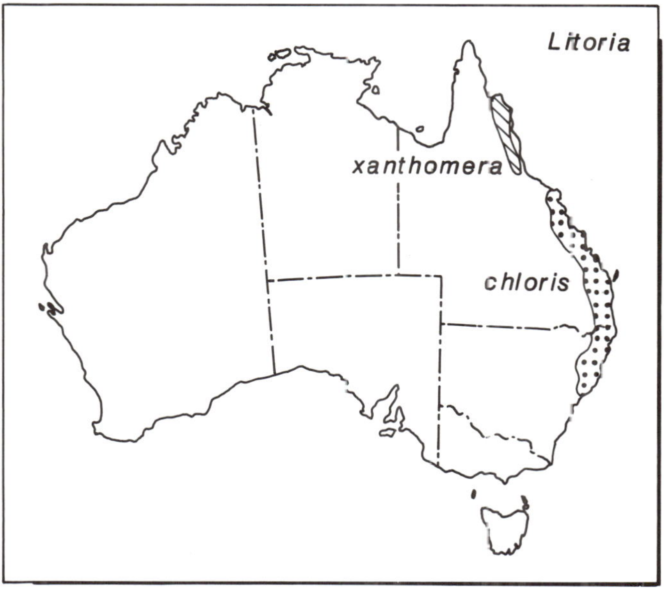

Prosperine. In spring and summer after heavy rain they congregate in and around flooded areas and mountain streams.

ADVERTISEMENT CALL: The call consists of a series of long "moans", usually followed by a soft trill. (Tape track 15).

BREEDING AND LARVAE: Males have been heard calling from October to February, during and after very heavy rain. Calling, amplexus, and oviposition occur in shallow pools in or beside rainforest streams. The eggs are laid singly or in small clumps, often entangled in the vegetation. Larvae are pale brown and reach a maximum length of 74 mm.

SIMILAR SPECIES: *Litoria chloris* and *L. xanthomera* resemble one another closely. They do not occur together — *L. chloris* ranges from northern New South Wales to Proserpine in Queensland, and *L. xanthomera* from Townsville to Cooktown. Posterior thigh colour distinguishes the species readily: blue/purple in *L. chloris* and orange in *L. xanthomera*.

Litoria gracilenta

(Peters)

Dainty Green Tree Frog, Slender Green Tree Frog

This is one of the most attractive of the green tree frogs. When resting, *L. gracilenta* adopts a typical cryptic pose, flattened against a leaf with the eyes closed and all markings on the sides of the legs and the body concealed. The dorsal surface is pale leaf green and the ventral surface pale yellow. The upper arm and dorsal surface of the thigh are straw yellow and the posterior of the thigh purple-brown. The tympanum is distinct and granular, and the iris bright orange. The hands and feet are fully webbed and the discs large.

SIZE: Males 31–42 mm; females 32–45 mm.

HABITAT: This species occurs along the Queensland–northern New South Wales coast. It may be found on reeds and floating vegetation in streams and swamps, but is equally common in disturbed areas such as banana plantations. During winter it has been found in hollow limbs in the crowns of *Eucalyptus tereticornis*, well away from water.

Litoria gracilenta, Bellingen, New South Wales *(Photo by Marion Anstis).*

Litoria xanthomera, Mt Spec, Queensland *(Photo by Steve Richards).*

ADVERTISEMENT CALL: A long drawn out "waaaaa", heard only in rainy periods.

BREEDING AND LARVAE: Eggs are deposited in November and attached to stems of grass in ponds. Tadpoles reach a maximum length of 34 mm and are dark brown to black. The tadpole life span is about fourteen weeks. The tooth formula is $\frac{1 \; {}^{1} \; 1}{1 \; {}_{2} \; 1}$.

SIMILAR SPECIES: Superficially *L. gracilenta* resembles *L. chloris* and *L. xanthomera*. It may be distinguished from both species by its size, its coarse, granular skin, and the colour of the posterior surfaces of the thighs: mauve-brown in *L. gracilenta*, blue/purple in *L. chloris* and orange in *L. xanthomera*.

Litoria xanthomera

Davies, McDonald and Adams

Orange-thighed Frog

This beautiful green tree frog is confined to coastal rainforest located between Cooktown and Townsville in north Queensland. It was distinguished from *L. chloris* as a valid species in 1986. In its overall proportions and appearance it is indistinguishable from the preceding description of *L. chloris* except for the colour of the back of the thighs (see below).

SIZE: Males 40–56 mm; females 43–55 mm.

HABITAT: Confined to dense rainforest where it may be encountered after heavy rain. *L. xanthomera* presumably lives in the tree canopy and descends to the ground only to breed.

ADVERTISEMENT CALL: As for *L. chloris*.

BREEDING AND LARVAE: Details of breeding are unknown. The tadpole is indistinguishable from that of *L. chloris* (see above).

SIMILAR SPECIES: This species may be distinguished from the closely related *L. chloris* by the colour of the posterior surface of the thighs: mauve to purple in *L. chloris* and orange in *L. xanthomera*. It should be noted that the species do not occur together. Hybrids have been produced in captivity and many have abnormalities of the skeleton, particularly dislocated hips.

Torrent Frogs (the *Litoria nannotis* Group)

The term "Torrent Frogs" is applied to a group of four species closely associated with fast flowing streams in rainforests on the eastern side of Cape York Peninsula of north Queensland. All of the members of the group produce only soft calls and *L. nannotis* and *L. lorica* lack vocal sacs. The nuptial pads bear black spines. The tadpoles have a flattened body and a large sucker-like mouth on the undersurface. There is great concern that some of these species may have joined the ranks of extinction as some have not been seen for two years or more (Ingram and McDonald 1994).

Key to the Torrent Frogs

(from Richards 1992)

1a. Snout short, nostrils on the end. Snout to vent length less than 40 mm 2
b. Snout rounded with nostril opening on the side. Size greater than 40 mm .... *Litoria nannotis* (p. 136)

2a. Males without black spines on the chest; tympanum indistinct; finger disc slightly expanded 3
b. Males with black spines on the chest; tympanum distinct; finger discs large .... *Litoria lorica* (p. 135)

3a. Fine spines on nuptial pad; ventral surface of body creamish .... *Litoria rheocola* (p. 140)
b. Coarse spines on nuptial pad; ventral surface of body pinkish .... *Litoria nyakalensis* (p. 138)

Litoria lorica

Davies and McDonald

Armoured Frog, Armoured Mist Frog

A small, grey or brown, dull coloured species with half-webbed fingers. Males have a protuberant prepollex on the thumb, bearing a nuptial pad of very large, black spines. The toes are fully webbed and fingers and toes bear large, terminal discs.

SIZE: Males 30–32 mm; females 33–37 mm.

HABITAT: Closely associated with rocks in fast-flowing streams in notophyll forest.

ADVERTISEMENT CALL: Unknown.

BREEDING AND LARVAE: Unknown. It is likely that the tadpole will resemble other members of this group in having a flattened body and large sucker mouth.

SIMILAR SPECIES: The secondary sexual character of black spines on the chest is not shared by either of the torrent species of similar size (*L. nyakalensis* and *L. rheocola*), whilst *L. nannotis* (which possesses spines) is substantially larger.

Litoria nannotis

(Andersson)

Torrent Tree Frog, Waterfall Frog

A torrential waterfall would seem the most unlikely habitat for frogs. Nevertheless, *L. nannotis* spends its life, both larval and adult, clinging to the slippery rocks of rushing waterfalls and in fast-flowing streams. It is a medium-sized species with a very broad, rounded snout. The dorsum is warty and grey to green with a network of heavy black markings. The ventral surface is pale and granular. The tympanum is indistinct, an unusual feature for this genus. Fingers and toes have very large discs. Fingers are less than half webbed, but toes are fully webbed. The thumb is only rudimentary but in breeding males it bears a large spiny nuptial pad. Spines also cover the chest region of males.

SIZE: Males 40–48 mm; females 49–55 mm.

HABITAT: Found in areas of rainforest or wet sclerophyll forest, usually restricted to areas near to waterfalls and cascades. May aggregate amongst rocks.

ADVERTISEMENT CALL: *Litoria nannotis* lacks a vocal sac and its call is a short, harsh growl.

BREEDING AND LARVAE: Very large cream eggs are laid in tough gelatinous masses under rocks in the stream bed. The colour of the tadpoles varies with the substrate. They have a sucker-mouth on the ventral surface, a strong muscular tail, and the tooth formula is $\frac{2}{3}$.

SIMILAR SPECIES: Amongst the *L. nannotis* group, *L. nannotis* is by far the largest and it is also distinguished readily by the sharply spinous form of the male nuptial pad.

toria lorica, Windsor Tableland, north Queensland *(Photo by John Wombey).*

itoria nannotis, Davies Creek Road, north Queensland *(Photo by Gordon Grigg).*

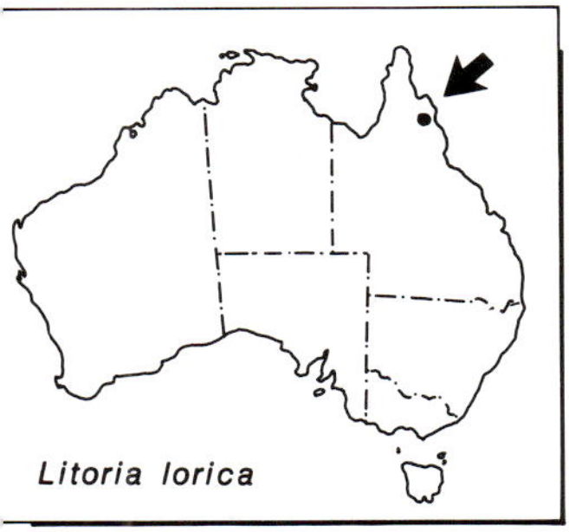

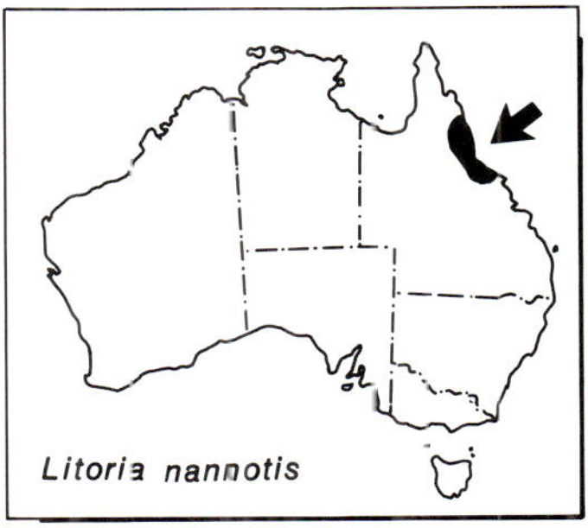

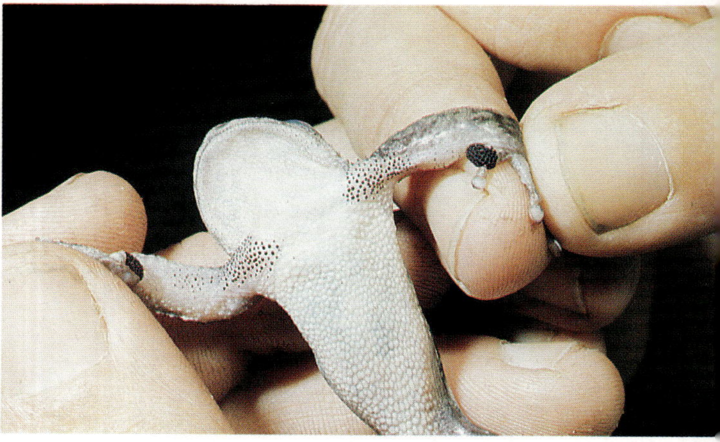

Litoria nannotis, showing nuptial pads and chest spines, Davies Creek Roa• north Queensland *(Photo by Gordon Grigg)*.

Litoria nyakalensis

Liem

Nyakala Frog, Mountain Mist Frog

A moderate sized species with a rounded snout, half-webbed fingers and extensively webbed toes. The forearms of the male are muscular and the nuptial pad is large and covered with large, black spines. The tympanum is small and partly hidden beneath the skin. In colour this species ranges from slate to dull brown, cream beneath, with a pinkish suffusion.

SIZE: Males 30–33 mm; females unknown.

HABITAT: Closely associated with streams in rainforest and wet sclerophyll forest. Frogs have been found on emergent rocks and boulders.

ADVERTISEMENT CALL: A single, soft rasping note repeated at regular intervals.

BREEDING AND LARVAE: The nature of the spawn is unknown. Tadpoles have a large sucker mouth on the undersurface of the head, a flattened head and body, and low fins. The body is light brown and the tail bears a number of discrete patches of dark pigment. Tadpoles can burrow into loose sand beneath rocks on the stream bed. The tooth formula is $\frac{2}{3}$.

SIMILAR SPECIES: See account of *Litoria rheocola*.

Litoria nyakalensis, Mt Lewis, north Queensland *(Photo by Martin Cohen).*

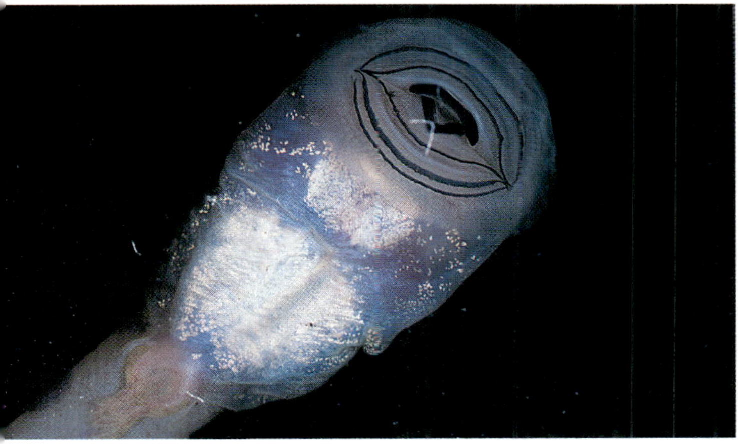

Litoria nyakalensis, tadpole mouthparts, Carbine Tableland, north Queensland *(Photo by Steve Richards).*

Litoria rheocola, Carbine Tableland, north Queensland *(Photo by Steve Richards)*.

Litoria rheocola

Liem

Creek Frog, Common Mist Frog

A moderate sized species with a sharply-defined canthus rostralis, small tympanum, half webbed fingers, fully webbed toes, and a small nuptial pad of a few separated spines on the thumb. A very dull coloured species, principally slate or tan above and dull creamish beneath, with a dark patch beginning between the eyes and narrowing or ending abruptly mid dorsally. Sometimes there is a greenish tinge on the flanks.

SIZE: Males 29–32 mm; females 33–38 mm.

HABITAT: Known only from fast-flowing streams in rainforest and in wet sclerophyll forest, from sea-level to 1 200 metres.

ADVERTISEMENT CALL: Males call from rocks or boulders in the streams or from adjacent vegetation. The call is a long, single "wreek".

BREEDING AND LARVAE: The cream eggs are laid in compact clumps on submerged rock surfaces. The tadpole attains a total length of 30 mm and is sandy to dark brown depending upon the substrate.

SIMILAR SPECIES: Externally *L. rheocola* and *L. nyakalensis* are similar but they can be distinguished most simply by the ventral coloration, *L. rheocola* lacking the pinkish tinge exhibited by the latter species.

Other *Litoria* Species

Each of the following species cannot be associated readily with any of the previous groups, or with one another.

Key to Other Species of *Litoria*

1a. White stripe from nostril to groin and a broader dark brown stripe from eye to groin; confined to Western Australia .... *Litoria adelaidensis* (p. 142)
b. White stripe absent, brown stripe present or absent; not found in Western Australia 2

2a. Small species (less than 25 mm) 3
b. Moderate or large species (greater than 25 mm) 4

3a. Body flattened, toes fully webbed, discs transversely oval .... *Litoria meiriana* (p. 149)
b. Body cylindrical, toes partially webbed, discs circular .... *Litoria microbelos* (p. 149)

4a. Outer metatarsal tubercle shovel-shaped (should also be compared with *Cyclorana* species) *Litoria alboguttata* (p. 142)
b. Outer metatarsal tubercle not shovel-shaped 5

5a. Confined to northern Australia 6
b. Confined to southeastern Australia 7

6a. Toes fully webbed; inhabits northern Western Australia and the northern portion of the Northern Territory .... *Litoria coplandi* (p. 145)
b. Toes one-half webbed; inhabits eastern Cape York Peninsula of Queensland .... *Litoria longirostris* (p. 147)

7a. Fingers slender and unwebbed; confined to New England Tableland of New South Wales ..... *Litoria piperata* (p. 150)
b. Fingers slightly webbed 8

8a. Confined to Victoria .... *L. spenceri* (p. 152)
b. Confined to Tasmania .... *L. burrowsae* (p. 144)

Litoria adelaidensis

(Gray)

Slender Tree Frog

This is one of only three hylids found in the south-west of Western Australia. The snout is distinctly pointed, the body long and slender. The dorsum is pale brown, green or brown with large green patches. The sides of the head and body are very dark brown or black, bounded beneath by a narrow white stripe. The belly is white. The posteriors of the thighs bear brilliant small red spots. The tympanum is quite distinct. Fingers have only a trace of webbing and large discs, toes are three-quarters webbed.

SIZE: Males 34–45 mm; females 42–47 mm.

HABITAT: Permanent swamps and lagoons in the wetter parts of the south-west of Western Australia. The frogs are usually found on the ground or among emergent vegetation at the water's edge.

ADVERTISEMENT CALL: A harsh, grating screech.

BREEDING AND LARVAE: Breeding commences in early spring. Eggs are deposited in loose masses attached to submerged vegetation at the surface of the water. The tadpoles are dark brown with a pair of pale stripes along the body.

SIMILAR SPECIES: No confusion is likely with any other species in the area.

Litoria alboguttata

(Günther)

Striped Burrowing Frog

Litoria alboguttata was at one time considered to be a member of the Leptodactylidae (Myobatrachidae). It has numerous close affinities with *Cyclorana,* but is considered a species of *Litoria* because of the internal structure of the ends of the fingers. *Litoria alboguttata* is the only species of *Litoria* that can burrow. The back is olive to pale green with brown patches and a pale yellow/green mid-vertebral stripe. The ventral surface

Litoria adelaidensis, Albany, Western Australia, *(Photo by Gordon Grigg) (Specimen courtesy Grahame Webb).*

Litoria alboguttata, Narrabri, New South Wales *(Photo by Gordon Grigg) (Specimen courtesy Ben Miller).*

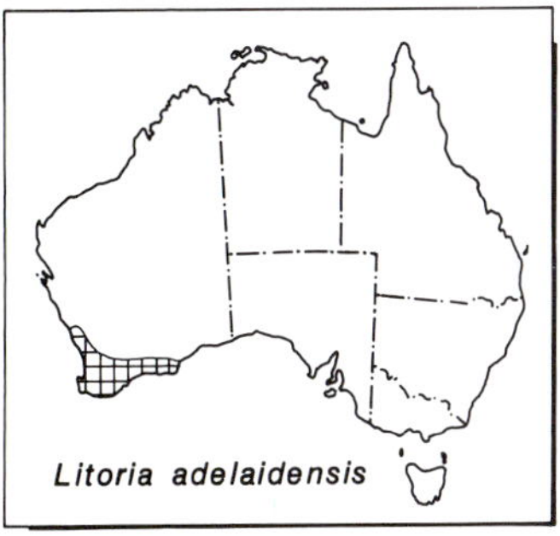

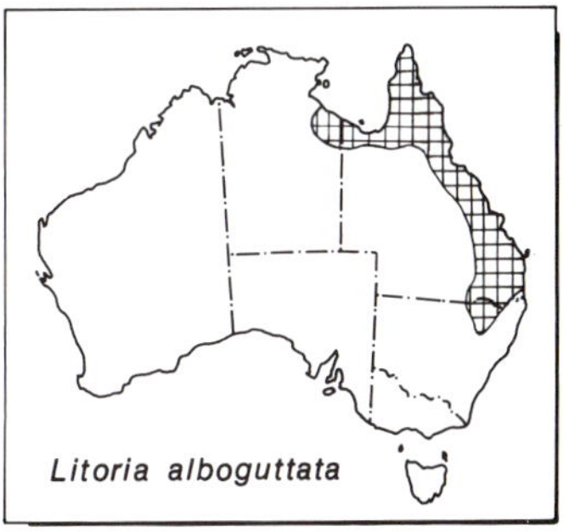

is dull white with darker mottling, particularly upon the throat. A distinctive feature is the dorso-lateral fold from the shoulder to the groin.

SIZE: Males 54–67 mm; females 61–83 mm.

HABITAT: Temporarily inundated grasslands and open forest.

ADVERTISEMENT CALL: A loud and rapid quacking.

BREEDING AND LARVAE: Unknown.

SIMILAR SPECIES: The burrowing habit and form of *L. alboguttata* are shared with *Cyclorana* species such as *C. australis*. The distinguishing feature of *L. alboguttata* is a series of short, elongated skin folds upon the back, whereas *C. australis* has a pair of long, continuous folds along the dorso-lateral surface.

Litoria burrowsae

(Scott)

Tasmanian Tree Frog

Litoria burrowsae is the only hylid species confined to Tasmania where it is restricted to montane areas. It is a robust species with long, fringed and basally webbed fingers and toes about three-quarters webbed. The coloration is highly variable from mottled brown and tan to green.

SIZE: Males 48–53 mm; females 50–60 mm.

HABITAT: Montane grassland and forests.

ADVERTISEMENT CALL: A very loud and penetrating honking.

BREEDING AND LARVAE: Spawn is attached in small clumps to vegetation in static or slowly flowing water. The tadpoles are a light yellow-brown. The tadpole tooth formula is $\frac{2}{3}$.

SIMILAR SPECIES: The only other hylid species found in Tasmania are *L. ewingii* and *L. raniformis*, neither of which is similar to *L. burrowsae* in external appearance.

Litoria burrowsae Arthur River, Tasmania *(Photo by Ron Milledge).*

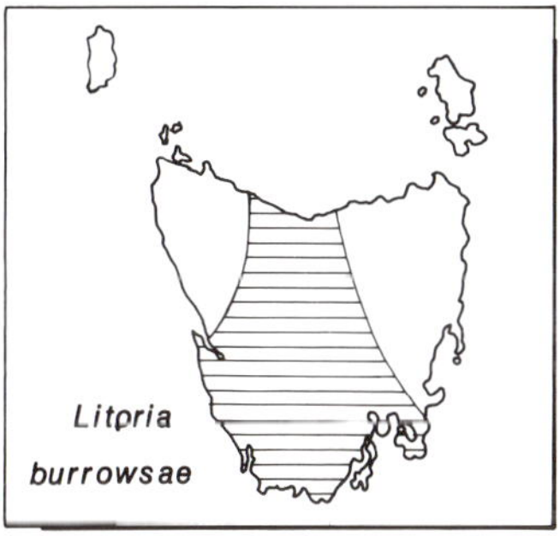

Litoria coplandi

(Tyler)

Copland's Rock Frog

This species is unusual because of its broad head and body and its habit of living on rock faces near escarpments. The fingers are long, slender, unwebbed and with large discs, whilst the toes have large discs and are fully webbed. The colour ranges from dull slate to brown, with indistinct patches of lighter and darker pigment.

SIZE: Males 29–36 mm; females 39–42 mm.

HABITAT: Exposed areas of rock, scree slopes and streams in exposed areas with rocky floors. Found aggregating in caves in the dry season.

Litoria coplandi Lake Argyle, Western Australia *(Photo by Bohdan Stankowich-Janusch).*

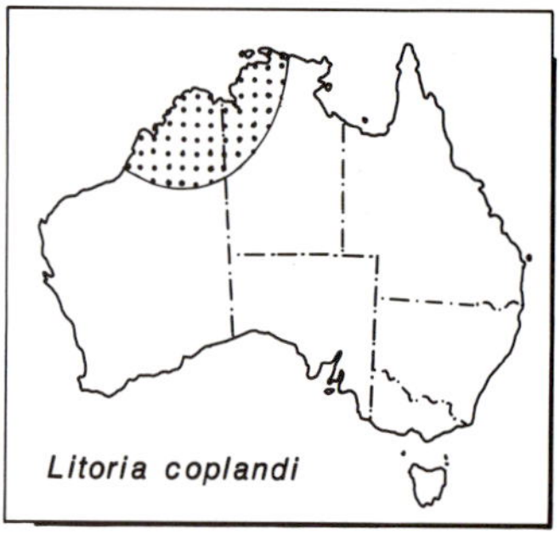

ADVERTISEMENT CALL: Unknown.

BREEDING AND LARVAE: Eggs are laid singly or in small clumps in rock pools where they sink to the floor. The tadpoles are dark brown with heavy pigmentation on the tail. Lifespan of tadpoles is at least 52 days. Mouthparts $\frac{2}{3}$.

SIMILAR SPECIES: On the isolated areas of rock where *L. coplandi* is found, only *L. personata* near Jabiru might be confused with it. The latter species can be distinguished by the broad stripe along the side of the body.

Litoria longirostris

Tyler and Davies

Long-snouted Frog, Scrub Rocket Frog

The least known of all the Australian hylid frog species, the only information about *L. longirostris* is derived from the preserved type series and the recent observations of McDonald and Storch (1993). The head is pointed and the canthus rostralis strongly curved. The body is slender and elongate. The fingers are long, slender and unwebbed and the discs large and oval. The toes are approximately one-half webbed. In preservation the overall colour is dull brown to dark grey.

SIZE: Males 25–27 mm; females 25–26 mm.

HABITAT: Known only from monsoon closed forest at Rocky River in the McIlwraith Range of the eastern Cape York Peninsula of northern Queensland. The frogs were found on the buttress of a tree and on rocks adjacent to a creek and on rocks in the stream.

ADVERTISEMENT CALL: Unknown.

BREEDING AND LARVAE: Clumps of 29–60 lime green eggs are laid out of water in September. The clumps have been found on tree trunks, upon rocks and under leaves near gently flowing or static water.

COMMENT: Green eggs are characteristic of frogs of the New Guinea *Litoria iris* group. The colour is believed to be due to the presence of the bile pigment biliverdin.

SIMILAR SPECIES: There are no species in Australia with which *L. longirostris* could be confused. Its most closely related species is probably *L. timida* of southern New Guinea.

Litoria longirostris, McIlwraith Range, north Queensland, *(Photo by Margar Davies) (Specimen supplied by Keith McDonald).*

Litoria meiriana, Katherine Gorge, Northern Territory *(Photo by Glenn Hoye)*

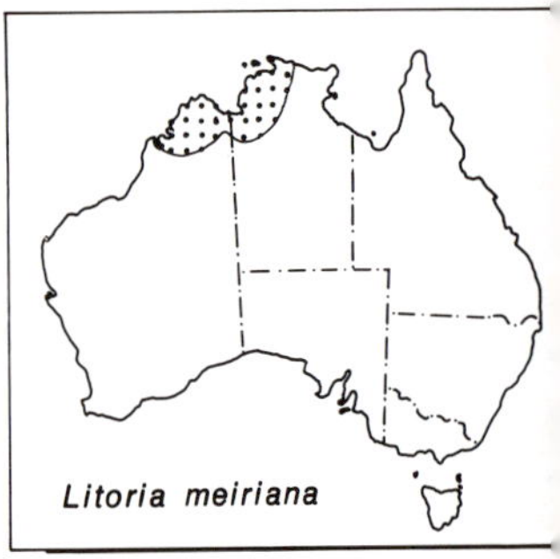

Litoria meiriana

(Tyler)

Rockhole Frog

This small frog can "skip" across the surface of water in a series of high-speed bounces. It is active in the daytime. There are two colour forms; the commonest is dull slate to purple with darker mottlings. The other form is dark brown above with pale dorso-lateral stripes on the sides of the head and body. The fingers are unwebbed and have large, transversely oval (not circular) discs on their tips. The toes are fully webbed.

SIZE: Males 16–20 mm; females 18–21 mm.

HABITAT: Closely associated with small pools upon the face of rock outcrops.

ADVERTISEMENT CALL: Unknown.

BREEDING AND LARVAE: Eggs are laid in small clumps attached to rocks or to the floor of temporary pools. The tadpoles are pale brown and attain a maximum length of 26 mm. Development can be completed in four weeks. The tadpole tooth formula is $\frac{2}{3}$.

SIMILAR SPECIES: Although *Litoria meiriana* is within the size range of a number of other northern Australian frogs, it can be distinguished readily by its transversely oval finger and toe discs, and its capacity to skip across the surface of water.

Litoria microbelos

(Cogger)

Javelin Frog

Litoria microbelos is the smallest of the Australian hylids, growing to a diminutive 16 mm. The dorsum is brown with a central portion somewhat lighter than the lateral areas. The ventral surface is variable off-white to light brown, with some darker speckling. There may be a few warts on the back and head, particularly above the eyes. The snout is long and sharply pointed. The fingers are free from web and the toes have only a trace of webbing.

SIZE: Males 14–16 mm, females 15–16 mm.

HABITAT: Known from the wetter areas of northern and central coastal Queensland, the Northern Territory and Western Australia, *L. microbelos* is found among long grass surrounding swamps, marshy areas and ponds.

ADVERTISEMENT CALL: A high-pitched buzzing.

BREEDING AND LARVAE: Not known.

SIMILAR SPECIES: *Litoria microbelos* is a miniature ground-dwelling hylid which can be recognized immediately by its size and by the amount of webbing between the toes.

Litoria piperata

Tyler and Davies

Peppered Tree Frog

This is a slender and delicate tree frog with expanded discs on the tips of the fingers and toes, no webbing between the fingers and fully webbed toes. The back is a diffuse mixture of pale green and dark areas with a mottling of slate and grey lacking any distinct pattern but with many flattened tubercles.

SIZE: Males 20–27 mm, females 24–31 mm.

HABITAT: Known only from dense vegetation close to water, large rocks and boulders of fast-flowing streams of the upper Clarence River catchment.

ADVERTISEMENT CALL: A series of short "chuck-chuck-chuck" notes.

BREEDING AND LARVAE: Unknown.

SIMILAR SPECIES: Some forms are somewhat similar to *L. phyllochroa* but can be distinguished by the warty, patterned dorsum and arms. It is also very similar to some colour forms of *L. spenceri* but the two species are separated by a considerable distance geographically and differ in body size.

itoria microbelos, Maningrida, Northern Territory *(Photo by Gordon Grigg).*

itoria piperata, Glen Innes, New South Wales *(Photo by Marion Anstis).*

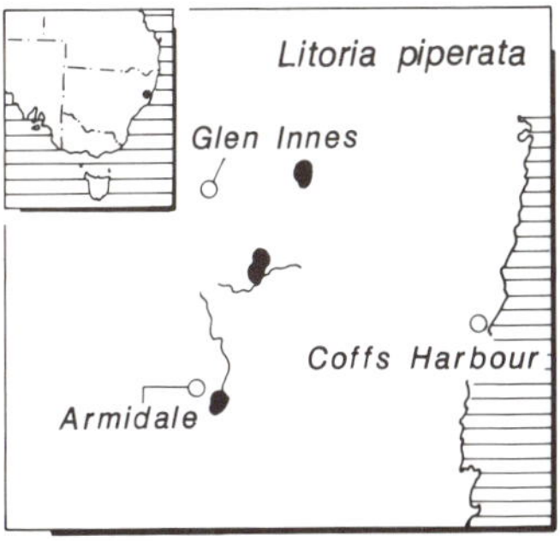

Litoria spenceri

Dubois

Spotted Tree Frog

Although described as the species *maculata* by Professor Baldwin Spencer in 1901, Alain Dubois in 1984 demonstrated that the name had been used previously for another species, and he substituted the name *spenceri*. *Litoria spenceri* is an endangered species. After its description from a specimen found at Powong, Victoria, it was not seen again until 1963 when Copland found three more specimens in Victoria. Since then more have been found in the extreme south of New South Wales at Bogong Creek and also in Victoria, and the Australian Capital Territory. The dorsal colour is quite variable in browns and greens. Darker markings in a reticulated pattern on a green background may be covered with minute pits. The ventral surface is pale white or yellow and granular. The groin and hindside of the leg are orange. The head is broad and a fold is present above the tympanum. A most unusual feature is the indistinct tympanum. The vomerine teeth are small and posterior to the choanae. Discs on fingers and toes are moderately developed with the fingers slightly and the toes fully webbed.

SIZE: Males 24–41 mm; females 37–52 mm.

HABITAT: All specimens collected so far are from mountainous areas with rocky flowing streams. Adults live amongst the rocks whereas juveniles have been found along shingle banks.

ADVERTISEMENT CALL: A series of 6–7 clear notes over a period of 3 seconds.

BREEDING AND LARVAE: Breeds in early summer in shallow streamside pools. Tadpoles undescribed.

SIMILAR SPECIES: *Litoria spenceri* is unlikely to be confused with any other species in its geographic range except for some forms of *L. verreauxii alpina*. *Litoria spenceri* is larger and the mottled colour pattern (bold blotches of green and brown in *L. v. verreauxii*) should easily separate the two species.

itoria spenceri, Mt Buffalo, Victoria *(Photo by Peter Robertson).*

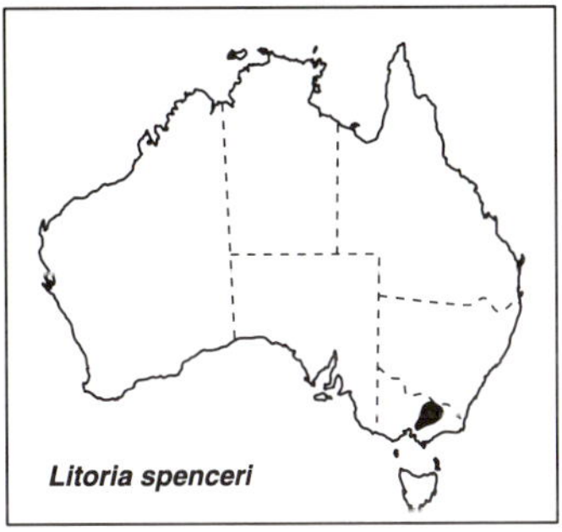

Genus *Cyclorana* Steindachner

The members of *Cyclorana* are burrowers. They have squat bodies, short, but muscular limbs and an enlarged inner metatarsal tubercle. Placed originally with the Myobatrachids, they are now considered members of the Hylidae because of similarities in chromosomes, muscle structure and tadpole morphology. In most species the eyes are large and protuberant and a dark stripe along the side of the head flanking the eye tends to create a fierce appearance. An exception is *C. platycephala* which has particularly small eyes and lacks a head stripe. It also differs in having extensive fleshy webbing between the toes, whereas other species have no more than one-half webbing.

Cyclorana species are distributed widely across northern Australia, particularly in the more arid areas and occur in all states except Victoria and Tasmania. They all burrow during dry times, shedding layer upon layer of skin to form an entire impermeable waterproof cocoon to prevent desiccation. They are not encountered frequently, but sometimes emerge in large numbers following rain. The difficulties of travel in the generally remote areas during rainy periods make field observations difficult.

Key to *Cyclorana* species

1a. Toes unwebbed or slightly webbed; eyes large and conspicuous 2
b. Toes fully webbed; eyes small, head flattened .... *C. platycephala* (p. 169)

2a. Adult size greater than 32 mm 3
b. Adult size less than 32 mm — Cape York Peninsula .... *C. manya* (p. 166)

3a. Tympanum clearly visible 4
b. Tympanum hidden .... *C. cryptotis* (p. 158)

4a. Longitudinal skin folds on lateral surface of body 5
b. Lateral surface of body lacking longitudinal skin fold ... 6

5a. Back of thighs variegated .... *C. australis* (p. 155)
b. Back of thighs not variegated .. *C. novaehollandiae* (p. 167)

6a. Dorsal skin smooth 7
b. Dorsal skin warty or ridged 9

7a. Back patterned with brown or fawn or yellow 8
b. Back pattern with dark brown or pale brown or slate or grey .... *C. longipes* (p. 163)

8a. Confined to eastern Queensland .... *C. brevipes* (p. 158)
b. Confined to central Northern Territory and extreme western Queensland .... *C. maculosus* (p. 164)

9a. Dorsal skin with a few long folds .... *C. maini* (p. 164)
b. Dorsal skin with numerous short folds or warty 10

10a. Dorsal skin with numerous short folds .... *C. verrucosa* (p. 170)
b. Dorsal skin warty 11

11a. Males 38–41 mm; females 38–43 mm; central Northern Territory and western Queensland .... *C. cultripes* (p. 161)
b. Males 42–48 mm; females 44–48 mm; northern Western Australia and adjacent Northern Territory .... *C. vagitus* (p. 170)

There is considerable uncertainty about the identification of *australis*-like individuals from northern Queensland, and also frogs resembling *C. longipes* from the same area and from central Australia. Field studies of mating calls may clarify their identities.

Cyclorana australis

(Gray)

Giant Frog

This is one of the largest of our ground-dwelling frogs and in northern Australia it is by far the most commonly encountered. During the wet season it can be found by day, basking beside temporary pools. The dorsum varies from pale grey or more commonly brown, with or without small darker markings, through to a bright green. The head is very large and broad. Broad lateral and uninterrupted, dark brown or chocolate head stripes extend from the tip of the snout to the flanks. There is often a pale mid-dorsal stripe. The skin is generally smooth except for a pair of longitudinal folds on each side of the back, and a second pair upon the flanks. The hind side of the thigh is variegated. The fingers are free, and the toes have a slight web.

SIZE: Males 71–79 mm; females 71–105 mm.

HABITAT: Principally grasslands and open woodlands.

ADVERTISEMENT CALL: Short "unk" repeated in long sequences.

BREEDING AND LARVAE: Up to 7 000 eggs are laid in temporary ponds from December to February. The spawn clump floats initially but later sinks. The tadpoles are a metallic sandy colour, reaching a length of up to 70 mm, and they occur commonly in large shoals. They have been found in water up to 43°C. The tadpole tooth formula is $\frac{1\ {}^{1}\ 1}{1\ {}_{2}\ 1}$.

SIMILAR SPECIES: *Cyclorana australis* most closely resembles *C. novaehollandiae* from which is it distinguished by its shallower head and possession of a variegated pattern upon the back of the thighs. It has been postulated that *C. australis* and *C. novaehollandiae* may hybridize where the populations meet in northern Queensland.

Cyclorana australis, Darwin, Northern Territory *(Photo by Gordon Grigg)*.

Cyclorana australis, green colour variant, Jabiru, Northern Territory *(Photo b Steve Donnellan)*.

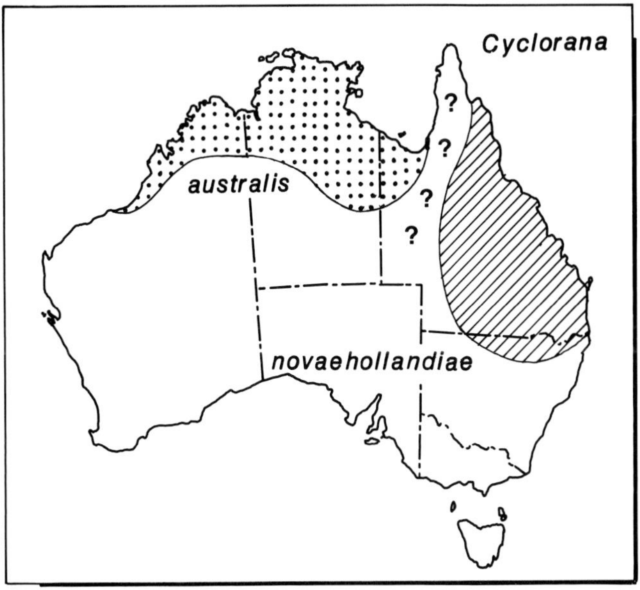
Cyclorana
australis
?
?
?
?
novaehollandiae

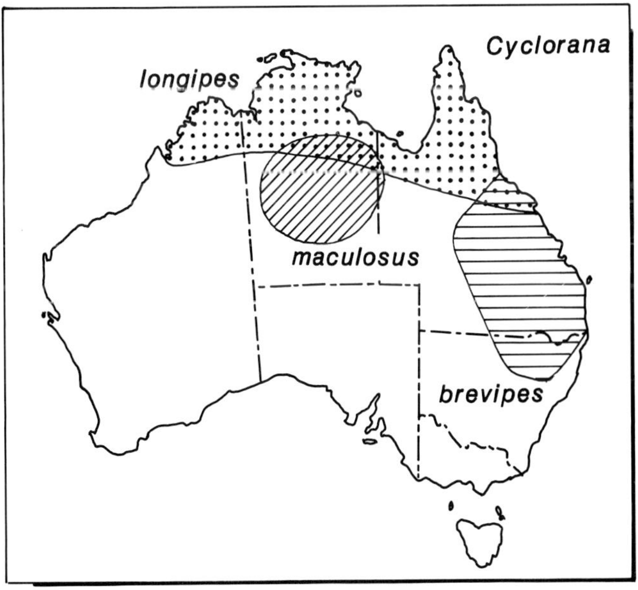
Cyclorana
longipes
maculosus
brevipes

Cyclorana brevipes

(Peters)

Short-footed frog

The dorsum of this compact frog is marked very beautifully with large irregular patches of dark brown pigment upon a pale sandy or yellowish brown background. The specific name alludes to the short feet. The ventral surface is white with some dark markings, particularly on the throat and a broad dark stripe on the side of the head runs from the snout to the shoulder. The tympanum is conspicuous but fairly small, the fingers are free and the toes slightly webbed.

SIZE: Males 36–45 mm; females 33–38 mm.

HABITAT: Open grassland and lightly forested areas.

ADVERTISEMENT CALL: A drawn out growl with a duration of 1.0–1.5 seconds.

BREEDING AND LARVAE: Males call from the margins of ponds. The tadpole and details of the life-history have not been reported.

SIMILAR SPECIES: *Cyclorana brevipes, C. longipes* and *C. maculosus* share irregular patterns of dark pigment upon a pale background. They are most readily distinguished by subtle differences in advertisement calls, but there remain uncertainties about the identity of specimens from several geographic areas.

Cyclorana cryptotis

Tyler and Martin

Hidden-ear Frog

The diffuse pattern of brilliant orange, tan or yellow, with dull brown markings often with a pale mid-dorsal stripe make this one of the most beautiful Australian species. It is small and stout, with short limbs and large protuberant eyes. The skin of the dorsum is finely shagreened and the tympanum is invisible. The fingers are unwebbed and the toes are approximately one-half webbed.

SIZE: Males 34–44 mm; females 36–46 mm.

Cyclorana brevipes, Inglewood, Queensland *(Photo by Gordon Grigg).*

Cyclorana cryptotis, Kununurra, Western Australia *(Photo by Bohdan Stankowich-Janusch).*

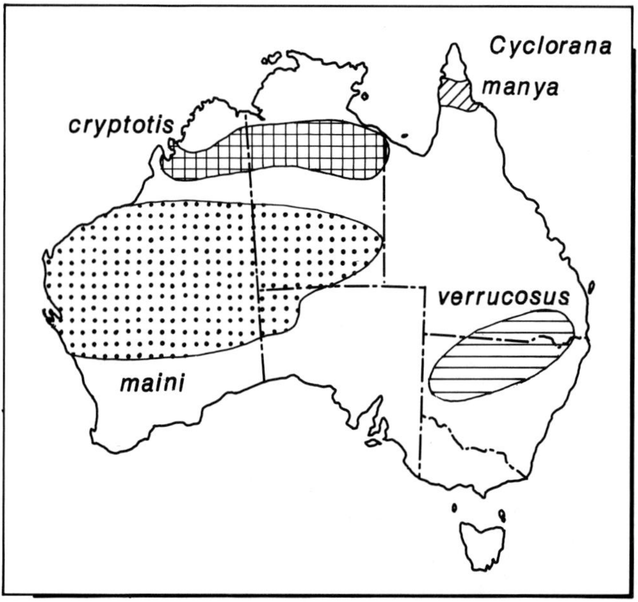
Cyclorana
manya
cryptotis
verrucosus
maini

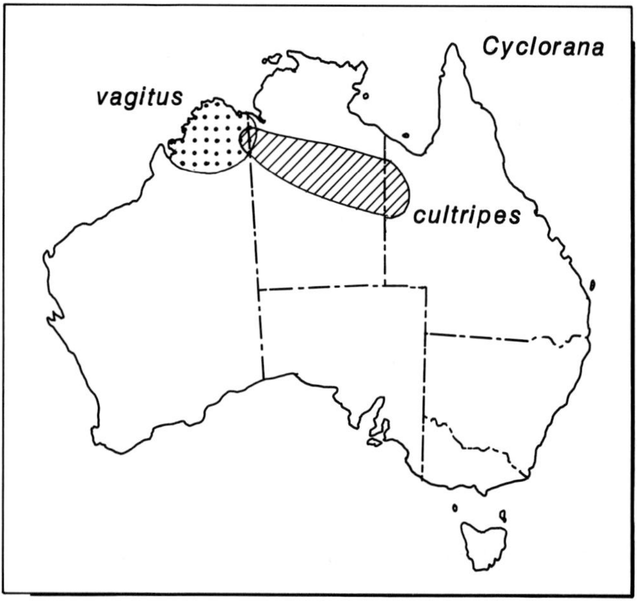
Cyclorana
vagitus
cultripes

HABITAT: Open grassland, usually associated with clay soils.

ADVERTISEMENT CALL: A loud, raucous bark which has a duration of one-third to one-half of a second, repeated frequently.

BREEDING AND LARVAE: Breeding takes place in January. Males call from the base of tall grasses at the edge of shallow flooded depressions. The water is usually milky. The tadpole is pale brown or cream and attains a maximum length of 46 mm. Larval life has been recorded at 24 days. The tadpole tooth row formula is $\frac{1\ {}^{1}\ 1}{1\ {}_{2}\ 1}$.

SIMILAR SPECIES: The striking coloration and hidden tympanum are features rendering this a distinctive species, unlikely to be mistaken for any other *Cyclorana*. There is a superficial resemblance to *Notaden melanoscaphus* but the latter can be distinguished readily by its black inner metatarsal tubercle (unpigmented in *C. cryptotis*).

Cyclorana cultripes

Parker

Knife-footed Frog

For many years any dull coloured *Cyclorana* was likely to be identified as *C. cultripes*. We now know that three species are involved, with minimal geographic overlap.

Cyclorana cultripes is usually dull grey, with a light, narrow mid-vertebral stripe, a broad, pale bar across the head behind the eyes and a broad, dull stripe on the side of the head. The ventral surface is white and granular and the throat has dark mottling. The skin is slightly warty. The tympanum is small but visible, the posterior of the thigh brown with white spots. The fingers are unwebbed and the toes about one-third webbed.

SIZE: Males 38–41 mm; females 38–43 mm.

HABITAT: Open grassland and sparsely wooded areas.

ADVERTISEMENT CALL: A harsh note with a duration of one-quarter of a second.

Cyclorana cultripes, Daly Waters, Northern Territory *(Photo by Margare Davies)*.

Cyclorana longipes, Derby, Western Australia *(Photo by Mike Mahony)*.

BREEDING AND LARVAE: Breeding takes place in January. Males congregate in flooded grassland and call from open situations at the edge of water. The form of the tadpole and details of development are unknown.

SIMILAR SPECIES: *Cyclorana cultripes* resembles most closely *C. maini*, *C. vagitus* and *C. verrucosus*. *Cyclorana maini* is pale brown rather than grey and is confined to mid-latitudes in Western Australia and the Northern Territory, whereas *C. cultripes* occurs further north. The geographic range of *C. vagitus* of the Kimberley overlaps with *C. cultripes* at the Western Australian/Northern Territory border, but the former is a larger species (males 42–48 mm; females 44–48 mm). *C. verrucosus* has an extremely warty dorsum (only slightly so in *C. cultripes*).

Cyclorana longipes

Tyler and Martin

Long-footed Frog

This species commonly shares the same environment as *C. australis*. The dorsum is usually pale grey with distinct slate patches, creating a striking appearance. There is a broad dark patch on the side of the head. The skin is smooth, the fingers unwebbed and the toes approximately one-half webbed.

SIZE: Males 38–46 mm; females 36–55 mm.

HABITAT: Open grassland.

ADVERTISEMENT CALL: Resembles the drawn-out lowing of cattle.

BREEDING AND LARVAE: Breeding takes place from December to February. Approximately 2 000 eggs are laid in loose clumps that sink before the early embryos emerge. The tadpoles are identical in shape and colour (metallic sandy) to those of *C. australis*, attaining a slightly smaller maximum length (60 mm). It is only in the final stages, when the characteristic colour pattern of the frog is developed, that the identity of the tadpole can be recognized. The tadpole tooth formula is $\frac{1\ ^{1}\ 1}{1\ _{2}\ 1}$.

SIMILAR SPECIES: This species is most likely to be confused with *C. maculosus* and *C. brevipes*. The former commonly exhibits a pale yellow background colour not observed in *C. longipes* and is slightly larger. *Cyclorana brevipes* is distinguished largely by call differences and appears to be geographically isolated.

Cyclorana maculosus

Tyler and Martin

Daly Waters Frog

There are many Australian species with unique restricted distributions. *Cyclorana maculosus* is one of them, for it is confined to an area from central Northern Territory to east of the Queensland border. It is an extremely beautiful species, commonly with a yellow background upon which are discrete patches of dark brown. The patches are irregular in shape and position, so that no two individuals are alike. The tympanum is conspicuous, the fingers are unwebbed and the toes are slightly less than one-half webbed. The skin is smooth.

SIZE: Males 34–55 mm; females 38–55 mm.

HABITAT: Open grassland.

ADVERTISEMENT CALL: A long moaning sound lasting approximately one-and-a-half seconds.

BREEDING AND LARVAE: Unknown. Probably breeds at the start of the northern wet season — in December.

SIMILAR SPECIES: Most likely to be confused with *C. brevipes* and *C. longipes*. Refer to account on p. 158.

Cyclorana maini

Tyler and Martin

Main's Frog

Named in honour of Professor A. R. (Bert) Main, this species thrives in some of the harshest country in central Australia. It is predominantly pale brown or dull green, bearing obscure and irregular patches or suffusions of a darker shade. A lateral head stripe is present or absent but never conspicuous. The tympanum is visible. The fingers are unwebbed and the toes slightly to one-half webbed. The skin is smooth or slightly warty.

yclorana maculosus, Daly Waters, Northern Territory *(Photo by Margaret Davies).*

yclorana maini, Hamersley, Western Australia *(Photo by Mike Mahony).*

SIZE: Males 35–46 mm; females 39–47 mm.

HABITAT: Open grassland, claypans, lightly forested country and temporary marshes.

ADVERTISEMENT CALL: The call is slightly less than one second in duration and resembles the bleating of sheep.

BREEDING AND LARVAE: An opportunistic breeder. Tadpoles are reported to resemble those of *C. platycephala* and are pink with an opalescent sheen in the latter stages of development.

SIMILAR SPECIES: The lack of a distinct pattern to the coloration is a feature shared with *C. cultripes* and *C. verrucosus*. The distribution of the latter is confined to the eastern half of the continent, separate from the western *C. maini*. *Cyclorana cultripes* has been confused with *C. maini* but tends to have a broad pale bar behind the eyes (lacking in *C. maini*) and distinct warts on the back.

Cyclorana manya

Van Beurden and McDonald

Small Frog

This species has the most restricted distribution of all *Cyclorana* species, known only from the central coastal portion of Cape York Peninsula. *Cyclorana manya* is also the smallest member of the genus. The colour of the living frogs has yet to be documented completely but is evidently a brown background with darker irregular markings upon it, and with a pale, narrow, mid-vertebral stripe. There is no dark lateral stripe on the side of the head. The tympanum is distinct. The fingers are unwebbed and the toes webbed only at the base.

SIZE: Males 27–30 mm; females 27–28 mm.

HABITAT: Open grassland.

ADVERTISEMENT CALL: A long call (1.1–1.3 seconds duration) described as a "plaintive bleat".

Cyclorana manya, Mitchell and Alice Rivers National Park, Cape York *(Photo by Steve Richards)*.

BREEDING AND LARVAE: Males have been found calling whilst floating in shallow temporary puddles. The habit of calling whilst floating is rare in *Cyclorana* and uncommon amongst members of other genera of Australian frogs. Spawn and tadpoles are unknown.

SIMILAR SPECIES: Van Beurden and McDonald (1980) compared the species with *C. brevipes* and *C. longipes*, perhaps because those are eastern Australian species. The diffuse nature of the colour pattern is more reminiscent of *C. maini*, but at present the diminutive size of *C. manya*, and its localized geographic distribution, provide a simple means of distinguishing it from other species.

Cyclorana novaehollandiae

Steindachner

New Holland Frog

This is a formidable frog; a mean machine, large and muscular, with a vast gape. The colour is not striking; a dense mottling of grey and brown, often with a pale mid-dorsal stripe. The head is massive and the eyes are perched high. The fingers are unwebbed and there is slight webbing between the toes.

Cyclorana novaehollandiae, Dorunda, Queensland *(Photo by Gordon Grigg)*.

Cyclorana platycephala, western Queensland *(Photo by Gordon Grigg)*.

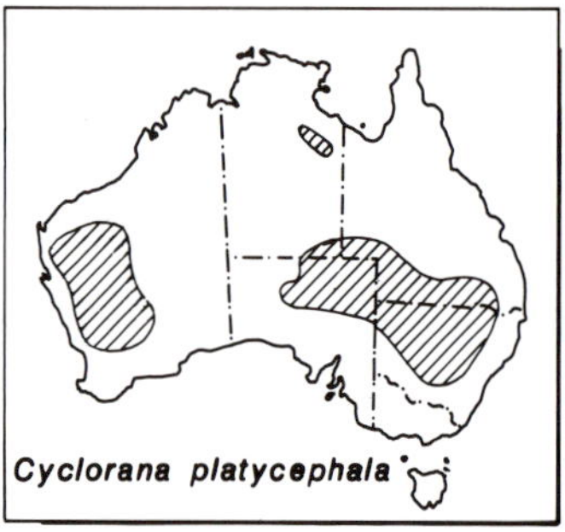

SIZE: Males 61–81 mm; females 71–101 mm.

HABITAT: Occurs throughout almost all of Queensland except at higher altitudes and in forested areas.

ADVERTISEMENT CALL: A repeated short explosive (one-quarter second) "unk".

BREEDING AND LARVAE: Eggs are deposited in clumps of about 1 000 in shallow, static or slowly-moving water. Details of development and of the form of the tadpole have not been published, although this species is reared in the USA as a vivarium animal.

SIMILAR SPECIES: Most closely related to *Cyclorana australis* and unlikely to be confused with any other. Distinguished by mottled dorsal coloration and thigh markings (see key), and by advertisement call. It is probable that the unusually broad-headed frogs in north Queensland are hybrids.

Cyclorana platycephala

(Günther)

Water-holding Frog

This is such a distinctive species that a good case could be made for placing it in a genus of its own. Its colour is unspectacular, ranging from a dull grey through dark brown to green, often with a pale mid-dorsal stripe and some dark irregular patches on the back. The eyes are very small and the head flattened. The tympanum is distinct and large in proportion to the size of the eye. The fingers are unwebbed and the toes fully webbed.

SIZE: Males 42–64 mm; females 50–72 mm.

HABITAT: Grasslands, temporary swamps, claypans and billabongs across diverse country.

ADVERTISEMENT CALL: A drawn out, snoring "maw-w-w-w . . . maw-w-w-w". Tape tracks 40, 45, 46, 47.

BREEDING AND LARVAE: Spawn is laid in large expanses of static water such as occurs during flooding. Tadpoles reach a maximum length of 60 mm. The tadpole tooth formula is $\frac{1\ ^{1}\ 1}{1\ _{2}\ 1}$.

SIMILAR SPECIES: This species has so many unique features that there is little chance of any confusion.

Cyclorana vagitus

Tyler, Davies and Martin

Wailing Frog

Cyclorana vagitus is one of several frogs which are confined to the mountainous Kimberley Division of Western Australia.

The colour is dull brown or tan, with obscure darker brown or green markings and, usually, a pale mid-vertebral stripe. The tympanum is conspicuous. The fingers are unwebbed and the toes slightly less than one-half webbed. The skin is consistently warty.

SIZE: Males 42–48 mm; females 44–48 mm.

HABITAT: Open grassland.

ADVERTISEMENT CALL: A regularly repeated note resembling the cry of a baby.

BREEDING AND LARVAE: Males normally call from the edge of water. During flooding near Derby, one was seen calling from a dried cattle pat floating on the surface. The eggs and tadpoles are unknown.

SIMILAR SPECIES: Most closely resembles *C. cultripes* which is smaller (males 38–41 mm; females 38–43 mm) and has a shorter call. The two species occupy a narrow zone of overlap at the Western Australia/Northern Territory border.

Cyclorana verrucosus

Tyler and Martin

Rough Frog

Confined to New South Wales and Queensland, this species bears a series of short skin folds arranged longitudinally upon its back. The back is marked irregularly and extensively with dark chocolate on a pale tan and dark green background. There is a dark stripe upon the side of the head. The tympanum is visible. The fingers are unwebbed and the toes webbed at the base.

SIZE: Males 35–45 mm; females 37–49 mm.

yclorana vagitus, Kununurra, Western Australia *(Photo by Bohdan Stankowich-Janusch).*

:yclorana verrucosus, Bingera, New South Wales *(Photo by Gordon Grigg).*

HABITAT: Open grassland.

ADVERTISEMENT CALL: Unknown.

BREEDING AND LARVAE: Unknown.

SIMILAR SPECIES: *Cyclorana cultripes* is the only eastern Australian species likely to be confused with *C. verrucosus*. However, it lacks the intense dark pattern of *C. verrucosus* and the back bears small warts, not skin folds.

FAMILY MYOBATRACHIDAE

Extant members of the family Myobatrachidae are confined to Australia. Related families are found in other southern continents of South America (Leptodactylidae) and Africa (Heleophrynidae). There are different opinions about the relationships between these groups, with the Australian species placed sometimes with the Leptodactylidae. It is convenient here to consider the Myobatrachidae as an endemic family with affinity to families on other continents.

Far greater diversity of habits and habitats is seen among the Australian Myobatrachidae than the Hylidae. The family includes species as diverse as those adapted specially for life in arid regions, through to high montane species with no tadpole stage, development confined to the egg capsule, and even parental care.

Key to Myobatrachidae

1a. Vomerine teeth present and distinctly visible 2
b. Vomerine teeth absent or, if present, poorly developed and indistinct 11
2a. Maxillary teeth present 3
b. Maxillary teeth absent. Body globose, snout short .... *Notaden* (p. 236)
3a. Tympanum distinctly visible 4
b. Tympanum indistinct or completely hidden 8
4a. Pupil either circular or a horizontal slit when contracted 5
b. Pupil a vertical slit when contracted .. *Lechriodus* (p. 344)
5a. Toes free from web 6
b. Toes with distinct web 7
6a. Tympanum large, its diameter equivalent to diameter of eye .... *Megistolotis* (p. 201)
b. Tympanum small, its diameter no more than one-half diameter of eye .... *Heleioporus* (p. 203)

7a. Stout burrowing frogs with short legs. No dark bars on hind limbs. Fleshy web deeply indented between toes. .... *Neobatrachus* (p. 212)
b. Large long-legged frogs. Dark bars on hind limb. Web penetrates outer metatarsals .... *Mixophyes* (p. 228)

8a. A pair of tusks at tip of lower jaw, large and distinct in males, much smaller in females .... *Adelotus* (p. 348)
b. No tusks in lower jaw 9

9a. Vomerine teeth in two broad series crossing the palate posterior to choanae .... *Limnodynastes* (p. 174)
b. Vomerine teeth in two short series between or slightly posterior to choanae 10

10a. Parotoid glands present and distinct. Restricted to Mount Baw Baw, Victoria .... *Philoria* (p. 242)
b. Parotoid glands absent. Northern New South Wales, Southern Queensland .... *Kyarranus* (p. 245)

11a. Tongue completely adherent to floor of mouth. Toes fully webbed .... *Rheobatrachus* (p. 350)
b. Tongue free behind. Toes not fully webbed 12

12a. Fingers and toes bearing distinct discs with median longitudinal groove .... *Taudactylus* (p. 250)
b. Fingers and toes lacking distinct discs with median longitudinal grooves 13

13a. Toes slightly dilated at tips. Males with lateral brood pouches opening posteriorly .... *Assa* (p. 346)
b. Toes not dilated at tips. Male lacks brood pouches 14

14a. Parotoid glands distinct .... *Uperoleia* (p. 312)
b. Parotoid glands poorly developed or absent 15

15a. Maxillary teeth absent 16
b. Maxillary teeth present 19

16a. Bright orange spot on undersurfaces and at insertion of limbs .... *Metacrinia* (p. 261)
b. No bright orange spots on undersurface of body 17

17a. Ventral surface of body heavily variegated with dark pigment .... *Pseudophryne* (p. 263)
b. Ventral surface of body unpigmented 18

18a. Head small; eyes minute. Body flattened and broadly oval. Arms short with short, broad fingers. Found south of Geraldton in WA .... *Myobatrachus* (p. 356)
b. Head small; eyes minute. Body globular. Arms and fingers not as above. Found north of Geraldton in Western Australia .... *Arenophryne* (p. 353)

19a. Narrow black stripe on side of head to forearm .... *Paracrinia* (p. 281)
b. No narrow black stripe 20

20a. Length of mature males to 35 mm. Vomerine teeth present. Belly smooth .... *Crinia* (p. 282)
b. Length of mature males to 25 mm. Vomerine teeth absent. Belly smooth .... *Geocrinia* (p. 302)

Genus *Limnodynastes* Fitzinger

The genus *Limnodynastes* occupies a wide variety of habitats and has a collective geographic range that includes most of the Australian mainland and Tasmania. One species *(L. ornatus)* occurs on several of the Torres Strait islands, whilst *L. convexiusculus* of northern and eastern Australia extends also to southern New Guinea. In all 12 species are recognized. The most recently described species *(L. depressus)* is known only from a single individual from a site now flooded beneath the man-made Lake Argyle in the Kimberley of Western Australia. The other species are relatively well known and, for many Australians, the conspicuous foamy nests of spawn of *Limnodynastes* floating on the water surface (so rare in foreign species) is the only kind of frog spawn that they have seen.

The size range of *Limnodynastes* species is from 29 mm in *L. ornatus* to 90 mm in *L. interioris*. Most species burrow or hide beneath ground cover as a shelter by day or as a longer retreat during dry periods. Several are equipped with spade-like inner metatarsal tubercles. The head is broad and the tympanum is not clearly defined. The vomerine teeth are unusual in being arranged in a long row behind the choanae.

The characteristic foam nest is constructed by the female who, in sequences of four to six strokes, raises her hands to the surface, trapping air bubbles which she then pushes downwards and backwards. The bubbles pass along her abdomen and between her legs, becoming embedded in the jelly surrounding the eggs whilst the eggs are being fertilized by the male. To facilitate the entrapment of the air bubbles the first two or three fingers of the female bear broad, lateral flanges of skin. Surprisingly, in southeastern South Australia, female *L. peronii* and *L. tasmaniensis* lack flanges on the fingers and do not construct foam nests. The lack of foam nest construction may be attributable to the unco-ordinated hand movements of the female — throwing most of the bubbles away from the emerging eggs.

Key to *Limnodynastes* species

1a. Inner metatarsal tubercle shovel-shaped, greater in length than first toe 2
b. Inner metatarsal tubercle not shovel-shaped, not longer than first toe 7

2a. A large oval gland on the upper surface of the tibia 3
b. No gland on the tibia 6

3a. Groin with scarlet markings 4
b. Groin without scarlet markings 5

4a. Dorsum usually with a pale mid-dorsal stripe. South-west of Western Australia .... *Limnodynastes dorsalis* (p. 178)
b. Dorsum usually with pale mid-dorsal stripe. Queensland and New South Wales .... *L. terraereginae* (p. 199)

5a. Toes one-third webbed. Ventral surface lemon-yellow .... *L. interioris* (p. 187)
b. Toes one-quarter webbed. Ventral surface white or heavily mottled with grey .... *L. dumerilii* (p. 181)

6a. Toes one-half to fully webbed .... *L. spenceri* (p. 196)
b. Toes one-quarter webbed .... *L. ornatus* (p. 189)

7a. Head flattened, upper eyelid projecting .... *L. depressus* (p. 178)
b. Head not flattened, upper eyelid not projecting 8

8a. First joint of first finger shorter than or equal to the second 9
b. First joint of first finger much longer than the second . 11

9a. First finger equal to or shorter than the second 10
b. First finger longer than second. A pink stripe over urostyle and on each flank .... *L. salmini* (p. 194)

10a. First finger equal to second. A single metatarsal tubercle. No mid-dorsal stripe .... *L. convexiusculus* (p. 176)
b. First finger shorter than second. Two metatarsal tubercles. Usually with mid-dorsal stripe .... *L. tasmaniensis* (p. 197)

11a. Toes with distinct webbing at base. Dorsum with irregular blotches. Usually no pale mid-dorsal stripe .... *L. fletcheri* (p. 186)
b. Toes long and free. Dorsum with light and dark stripes. A pale mid-dorsal stripe .... *L. peronii* (p. 191)

Limnodynastes convexiusculus

(Macleay)

Marbled Frog

This is a moderate to large-sized species with a fairly sharp snout and stout body. Exceptionally large individuals bearing sharp black spines over the entire back have been found in Darwin. The dorsum is warty and grey or olive with darker irregular blotches or a heavy network of dark markings. There is a black stripe on the side of the head, the belly is smooth and white, and the throat mottled in males. The fingers are free from webbing. The first finger is equal in length to the second. The toes are unwebbed and there is a prominent inner but no outer metatarsal tubercle.

SIZE: Males 46–50 mm; females 52–61 mm.

HABITAT: Usually found in swampy areas among long grass.

ADVERTISEMENT CALL: A loud, high-pitched "honk" repeated every couple of seconds and often continued for long periods.

BREEDING AND LARVAE: Breeds between October and March (earliest in the eastern portion of its range). Males call from concealed positions amongst grass and debris and even in flooded burrows vacated by crabs. Eggs are deposited in a small floating foam nest. Tadpoles are entirely black and attain a maximum length of 70 mm. The tadpole tooth formula is $\frac{5}{3}$.

SIMILAR SPECIES: It is only at Kununurra and at Newry Station to the west and in central coastal Queensland that *L. convexiusculus* occurs with *L. tasmaniensis*, with which it is most likely to be confused. Adults of the latter are smaller, with a smooth (not warty) back, and the calls are quite different.

mnodynastes convexiusculus, Cooktown, Queensland *(Photo by Gordon Grigg)*.

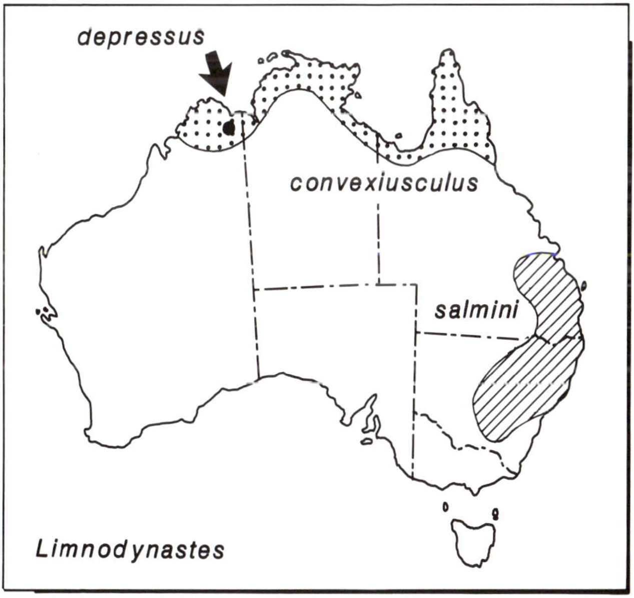

Limnodynastes depressus

Tyler

Flat-headed Frog

This species has been collected only at the original site of the Argyle Homestead which is now flooded beneath the man-made Lake Argyle. It was described in 1976 and intensive collecting at the northern end of the lake has failed to relocate it. The head is exceptionally flattened, the eye small and the eyelids jut outwards like a shelf. The tympanum is completely covered with skin. The skin is smooth. The fingers and toes are long, unfringed and unwebbed. In preservative the back is very pale grey, marked with numerous small irregular areas of brown.

SIZE: Known only from the type specimen: an adult male with a length of 39 mm.

HABITAT: Probably open grasslands near permanent water.

ADVERTISEMENT CALL: Unknown.

BREEDING AND LARVAE: Unknown.

SIMILAR SPECIES: Most similar to *L. convexiusculus*, from which it is distinguished by its flat head, large upper eyelids and smooth skin.

Limnodynastes dorsalis

(Gray)

Pobblebonk, Sand Frog or Bullfrog

This species, together with *L. dumerilii*, *L. interioris*, and *L. terraereginae*, is part of a group known collectively as the "*dorsalis* complex" because they have similarities in body shape, breeding biology, and mating call structure. In all members of this "complex", there is a pale glandular ridge from the eye to the corner of the mouth, and a large oval gland on the tibia.

Limnodynastes dorsalis is confined to the south-west of Western Australia and is a moderately large and stout frog. The dorsum is smooth, with large dark brown spots on a pale brown or olive background. A thin, pale mid-dorsal stripe is nearly always present. The belly is smooth and either pale yellow or white. The groin is a rich scarlet. The fingers are free from webbing,

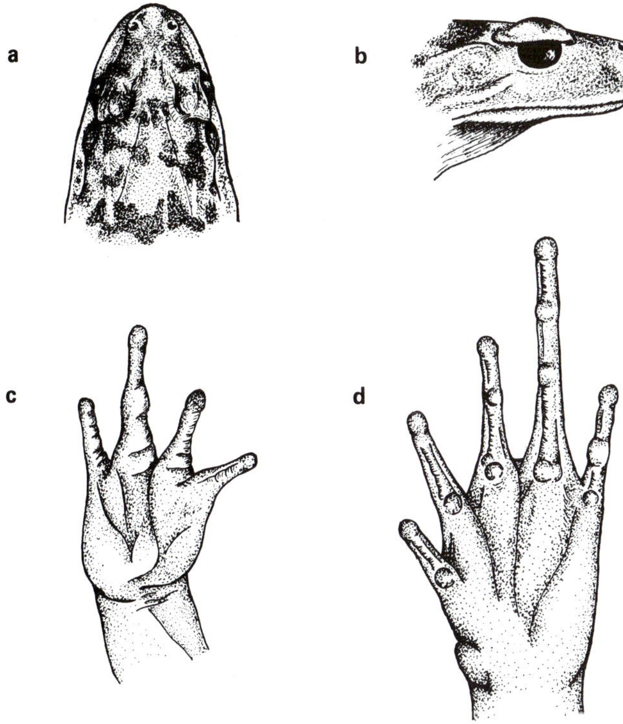

Limnodynastes depressus WAM R43896 (holotype) (a) dorsal view of head; (b) lateral view of head; (c) palmar view of hand; (d) plantar view of foot *(From Records of the Western Australian Museum, with permission).*

imnodynastes dorsalis, Albany, Western Australia *(Photo by Gordon Grigg) (Specimen courtesy Grahame Webb).*

and the relative finger lengths are 3 > 2 > 1 = 4. The toes are slightly webbed. The inner metatarsal tubercle is very large and it is longer than the first toe.

SIZE: Males 60–64 mm; females 66–73 mm.

HABITAT: *Limnodynastes dorsalis* is common in permanent dams, swamps and creeks in many areas of the south-west of Western Australia. Occasionally they are found foraging in the open, but more commonly they hide in thick vegetation by the bank.

ADVERTISEMENT CALL: The calls of all the *L. dorsalis* species group are superficially similar, consisting of a single short "bonk", which may be repeated rapidly.

BREEDING AND LARVAE: *Lymnodynastes dorsalis* breeds in winter. Males call from hidden sites, either under overhanging banks or in thick vegetation. The eggs are deposited in a large frothy mass which is usually attached to vegetation. Tadpoles reach a total length of 65 mm and have a dark body with a high, mottled tail fin. The tooth formula is $\frac{2\ ^{2}\ 2}{3}$.

SIMILAR SPECIES: No other *Limnodynastes* species occur in southwestern Australia. Any superficial resemblance to *Heleioporus barycragus* can be resolved by the absence of a tibial gland in that species.

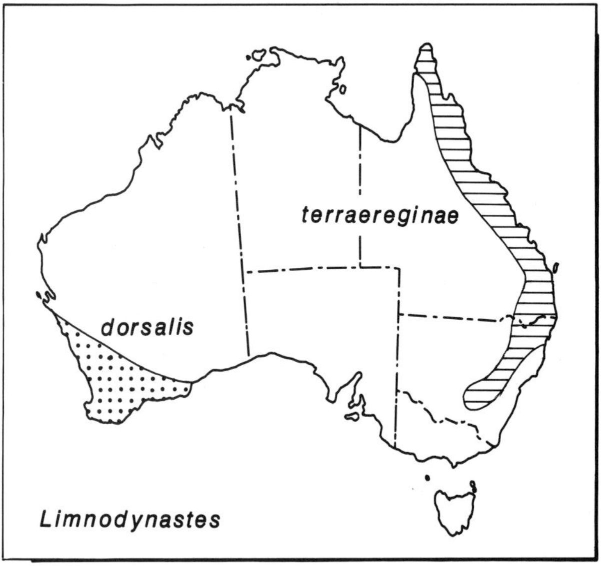

Limnodynastes dumerilii

(Peters)

Bullfrog or Pobblebonk

This species shows a great deal of variation in colour pattern, body length and mating call structure. Five sub-species have been recognized in southeastern Australia.

Limnodynastes dumerilii is a moderate to large stout frog with a very broad, rounded head and short thick limbs. The dorsum is rough and warty in texture and the colour varies from pale grey, through olive-green, to dark brown or black. A pale mid-dorsal stripe is usually present in eastern populations. In western and southern forms, the dorsum tends to be paler with irregular dark markings. The sides are often beautifully marked with a bronze and purple sheen with black mottling. The pale supralabial gland and the tibial gland are conspicuous and commonly gold in colour. There is only a slight web between the toes and none between the fingers. Finger lengths are $3 > 2 > 1 = 4$.

SIZE: Males 52–70 mm; females 52–83 mm.

HABITAT: This species burrows in loamy soils and forages on the surface after rain. It is found commonly in suburban gardens in southeastern Australia.

ADVERTISEMENT CALL: A single note varying from a musical banjo-like "bonk" to a harsher "tok", very similar to *L. peronii*. Several frogs often form a chorus, calling in sequence, each frog calling at a slightly different pitch, almost as if by design. (Tape tracks 6, 25, 38.)

BREEDING AND LARVAE: Males migrate up to 1 km to breeding sites: dams, small lakes, marshes and slow-moving streams. They call from hidden positions at the edge. Up to 3 900 eggs are laid in spring in a floating foam nest. The tadpoles are dark grey to black. In warm waters they complete development in 4–5 months; in cooler situations they can take 12–15 months. The tooth formula is $\frac{4\;^{1}\;4}{1\;_{2}\;1}$.

Limnodynastes dumerilii dumerilii, Glen Lyon, Queensland *(Photo by Gordon Grigg)*.

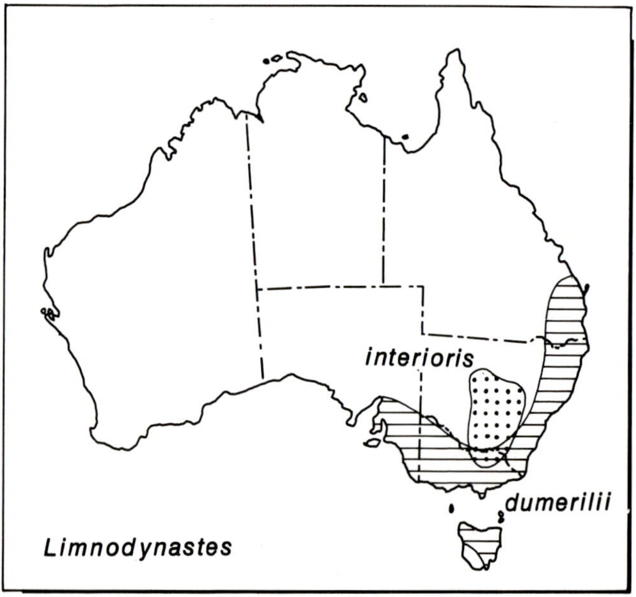

SIMILAR SPECIES: This species can be distinguished from *Heleioporus australiacus* by the conspicuous tibial gland and lack of a cone-shaped nuptial spine in males. *Limnodynastes dumerilii* can be distinguished from others in the *dorsalis* complex by the vestigial web (cf. *L. interioris*) and the lack of scarlet in the groin (*L. dorsalis* and *L. terraereginae*).

Key to sub-species of *Limnodynastes dumerilii* (from Martin 1972)

1a. Toes with only a trace of webbing 2
 b. Webbing on toes moderately developed and clearly visible 4

2a. Vertebral stripe present 3
 b. Vertebral stripe absent .... *L. dumerilii variegatus*

3a. Body length less than 50 mm; flanks suffused with lemon yellow .... *L. dumerilii grayi*
 b. Body length usually more than 50 mm; flanks not yellow .... *L. dumerilii insularis*

4a. Body length less than 70 mm .... *L. dumerilii dumerilii*
 b. Body length 70 mm or more .... *L. dumerilii fryi*

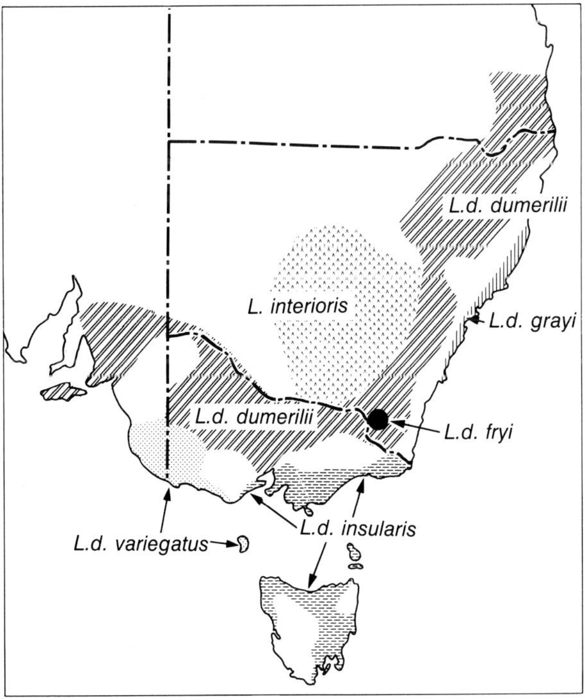

Limnodynastes dumerilii fryi, Thredbo, New South Wales *(Photo by Angus Martin).*

Limnodynastes dumerilii grayi, Putty, New South Wales *(Photo by Gordon Grigg).*

Limnodynastes dumerilii insularis, Frenchs Narrows, Victoria *(Photo by Gordon Grigg)*.

Limnodynastes dumerilii variegatus, Mortlake, Victoria *(Photo by Mike Mahony)*.

Limnodynastes fletcheri

Boulenger

Long-thumbed Frog, Barking Frog

This is a medium-sized frog which closely resembles *L. tasmaniensis* but which can be separated by the characteristics given in the previous description. The dark patches on the dorsum are more irregular in size and shape than the markings in *L. tasmaniensis*. A rose-coloured patch is usually present on each upper eyelid. The ventral surface is smooth and white. The snout is somewhat rounded and the tympanum is indistinct. There is definite webbing at the base of the toes and a small inner metatarsal tubercle. Webbing is not present on the fingers and the first finger is longer than the second (finger-lengths $3 > 1 > 2 = 4$).

SIZE: Males 37–46 mm; females 38–55 mm.

HABITAT: This species is usually found on the shores and banks of lakes and rivers. By day they hide under large rocks and logs and they have been found in cracks in dried mud. In dry weather they often aggregate in large groups.

ADVERTISEMENT CALL: A short barking "rok" or "whuck". (Tape track 36.)

BREEDING AND LARVAE: The breeding season of this frog varies with the amount of rainfall in a particular region. In the wetter parts of its range it breeds from October to March and, in the drier areas, at any time of the year after rain. In Queensland, breeding has been observed in winter. Tadpoles reach a maximum length of 68 mm. The tooth formula is $\begin{array}{c} 3 \\ 1 \quad 1 \\ \hline 1 \quad 1 \\ 2 \end{array}$.

SIMILAR SPECIES: See *L. tasmaniensis*.

Limnodynastes fletcheri, Inglewood, Queensland *(Photo by Gordon Grigg).*

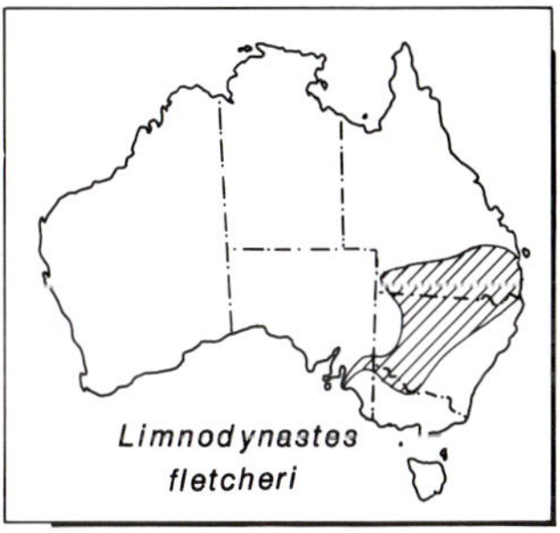

Limnodynastes interioris

(Fry)

Giant Bullfrog

Limnodynastes interioris is the largest and most colourful of the *dorsalis* complex. The dorsal surface is pale brown with a few darker spots. The flanks are marked with a wide, bright, copper-orange band with an irregular black band below it. The ventral surface is bright yellow. The vomerine teeth are similar to others in the complex. Supralabial and tibial glands are present. The toes are about one-third to one-half webbed and there is a large "shovel" on the foot.

SIZE: Males 70–90 mm; females 73–88 mm.

HABITAT: Inhabiting the drier generally sandy central portion of New South Wales and northern Victoria, *L. interioris* is an efficient burrower, spending the daylight hours and drier months buried beneath the surface. They have been recently found aestivating in the mounds of the Mallee Fowl in central New South Wales. Dams with marginal vegetation are the preferred breeding sites.

ADVERTISEMENT CALL: A very deep "thud" similar to *L. dumerilii*. (Tape track 39).

BREEDING AND LARVAE: In spring and summer males call while concealed in burrows in the banks of dams. Eggs are deposited in a frothy mass in the burrows. The tadpole tooth formula is $\frac{\begin{matrix}1\\5\quad 5\end{matrix}}{\begin{matrix}1\quad 1\\2\end{matrix}}$.

SIMILAR SPECIES: *Limnodynastes interioris* can be distinguished easily from others in the *dorsalis* complex by its brilliant lateral and ventral markings together with its large size and extensive webbing.

Limnodynastes interioris, Lake Cowal, New South Wales *(Photo by Gordon Grigg)*.

Limnodynastes ornatus

(Gray)

Ornate Burrowing Frog

Limnodynastes ornatus is a stout medium-sized frog with a broad and angular head and protruding eyes. There is enormous variation in the pattern of the dorsum. Commonly a pale sandy background colour is patterned intricately with darker markings, edged in black. There is often a distinct butterfly-shaped light patch behind the eyes. Small warts, often tipped in red, are scattered over the dorsum. The ventral surface is white and smooth. The vomerine teeth are to the rear of the choanae and the tympanum is indistinct. The limbs are pale grey, barred or spotted with darker markings, the fore-limbs are slim and the hind limbs are short and stout. The toes have a slight web and there is a very large shovel-shaped inner metatarsal tubercle which is slightly longer than the second toe. The fingers are devoid of web, relative lengths being $3 > 1 = 4$. The male nuptial pad occurs extensively on the first two or three fingers.

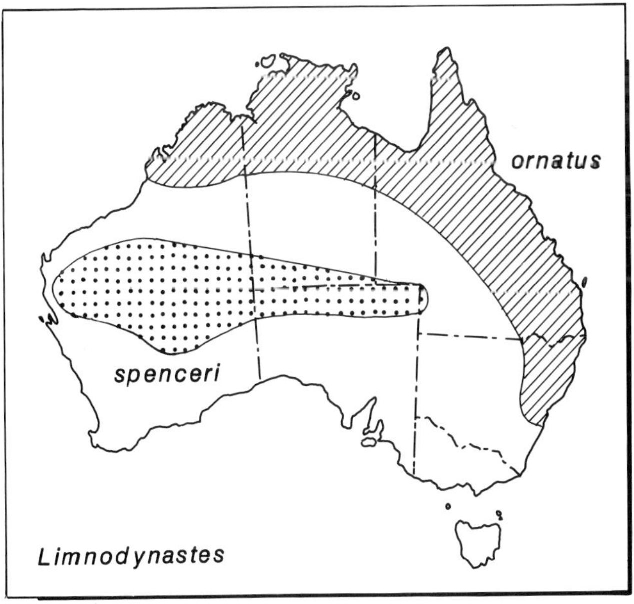

Limnodynastes ornatus, Cania Gorge, Queensland *(Photo by Gordon Grigg)*.

Limnodynastes ornatus. This species exhibits striking diversity in its markings. Pormpuraaw, Queensland *(Photo by Gordon Grigg)*.

SIZE: Males 29–37 mm; females 35–42 mm.

HABITAT: *Limnodynastes ornatus* is a versatile species, thought originally to inhabit only the drier regions, but now known from many coastal areas across the north of the continent. Often found in dry sandy watercourses some distance from permanent water, they burrow very efficiently and in arid country they are known to spend the dry season in a dormant state well below the surface.

ADVERTISEMENT CALL: A very short, nasal "unk" repeated slowly.

BREEDING AND LARVAE: This species breeds only after heavy rain. Males call while floating, spreadeagled, in the water, which may be only a small puddle. Up to 1 630 pigmented eggs are deposited in small foam masses which collapse rapidly to form a thin floating film if the water is warm. The tadpoles are pale brown. They can complete their development in as little as 21 days. Their tooth formula is $\frac{3\ ^{1}\ 3}{1\ _{2}\ 1}$.

SIMILAR SPECIES. *Limnodynastes ornatus* is closely related to *L. spenceri*, and can be distinguished by the reduced webbing and different mating call.

Limnodynastes peronii

(Duméril and Bibron)

Striped Marsh Frog, Brown Frog

This is one of the most common frogs in eastern Australia. The dorsal surface is marked attractively with a series of longitudinal dark and light brown stripes, and there is frequently a pale mid-dorsal stripe. The striped dorsal pattern breaks up laterally into a series of blotches. The ventral surface is white, except for the throat of the male, which has a yellow wash and dark brown mottling. The snout is rather pointed and the iris is golden above and dark brown below. Vomerine teeth are in two transverse series to the rear of the choanae, the tympanum is indistinct. The toes are very long and free from web, with a small inner metatarsal tubercle. The fingers are without webbing although most breeding females have prominent flanges. The forearms show sexual dimorphism: in breeding males they are huge and swollen and the tip of the first finger is sharp with a bone protruding through the skin to form a nuptial spine. Relative finger-lengths are $3 > 2 \geq 4 \geq 1$ and there is one large inner and two small outer metacarpal tubercles.

SIZE: Males 48–69 mm; females 46–73 mm.

HABITAT: This species is widespread and adaptable, occurring throughout the coast and ranges of eastern Australia. In suburban areas it is commonly found using outdoor fish ponds as breeding sites, annoying residents at night. *Limnodynastes peronii* is secretive by day, hiding under logs, stones, or leaf litter. It burrows effectively despite the lack of enlarged metatarsal tubercles.

ADVERTISEMENT CALL: Males call from concealed locations under overhanging ledges or from thick vegetation or forest debris. Whereas the casual calling heard by day is often made by frogs on land, at night they call from the water, often from concealed sites, floating at the surface with vocal sacs distended. The call is a soft explosive "whuck" similar to *Limnodynastes dorsalis*. Usually the call is a single note, but doublets and triplets have been heard. (Tape track 22.)

BREEDING AND LARVAE: Breeding occurs from August to March. 700–1 000 small unpigmented eggs are deposited in a foam mass entangled in vegetation at the edge of the water. In the lower south-east of South Australia females lack finger flanges and do not produce a foam nest. The tadpoles reach 65 mm in length and are pale brown with the adult dorsal pattern becoming apparent as the tadpole is developing the fore-limbs. The tadpole tooth formula is $\frac{2\;{}^{2}\;2}{1\;{}_{2}\;1}$.

SIMILAR SPECIES: *Limnodynastes peronii* could be confused with *L. salmini* or *L. tasmaniensis*. The last is much smaller (47 mm) and, although the body shape is essentially the same, the dorsal pattern is spotted not

mnodynastes peronii, Casino, New South Wales *(Photo by Gordon Grigg).*

mnodynastes peronii, early metamorph, Putty, New South Wales *(Photo by Gordon Grigg).*

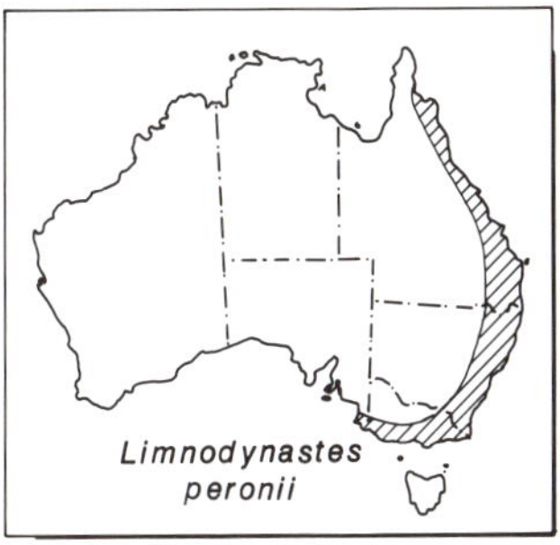

striped. As a further check, the relative finger-lengths of *L. tasmaniensis* are 3 > 2 > 4 > 1. *Limnodynastes salmini* can be distinguished by the presence of pink-orange dorsal and lateral stripes. Also the first finger is slightly longer than the second and it lacks the nuptial spine of *L. peronii*. In New South Wales, *L. peronii* and *L. salmini* are mostly isolated from each other, their ranges overlapping only in northern New South Wales and south-east Queensland.

Limnodynastes salmini

Steindachner

Salmon-striped Frog, Steindachner's Frog

This large and very colourful species has a body shape rather like *L. peronii* but it is usually stockier and has a rounder snout. The back is greyish-brown with a pattern of dark brown or black spots or blotches which sometimes may appear to be stripes rather than spots. Three brown to salmon-pink stripes are always present, one being mid-dorsal (the length of the urostyle), the others from the groin to the top of the shoulder. A light patch is often present just behind the head. The sides, groin, and anterior and posterior surfaces of the thigh are beautifully mottled in black and white. The ventral surface is smooth and white. The vomerine teeth are to the rear of the choanae and are very broad, while the tympanum is more distinct than in any other member of the genus. The toes are short, and there is a small round inner metatarsal tubercle. The fingers lack webbing and there is a small inner metacarpal tubercle which enlarges to a breeding pad. The finger lengths are 3 > 1 > 2 = 4.

SIZE: Males, 61–65 mm; females 68–76 mm.

HABITAT: This frog is found in a variety of situations, usually under logs, loose bark, or stones.

ADVERTISEMENT CALL: A resonant musical "wow". (Tape track 37.)

BREEDING AND LARVAE: Males call between September and April from matted vegetation on the edges of flooded depressions. Larvae are very similar to other *Limnodynastes* species. Further details of the life history are unknown.

SIMILAR SPECIES: See *L. peronii*.

nnodynastes salmini, Moura, Queensland *(Photo by Gordon Grigg) (Specimen courtesy Graeme Watson and Margaret Davies).*

nnodynastes spenceri, Lake Way, Western Australia *(Photo by John Wombey).*

Limnodynastes spenceri

Parker

Spencer's Burrowing Frog

Although this species has been known since its description in 1940 there has been no detailed comparison with *L. ornatus*, to which it is very closely related. It is generally distinguished by its having more extensive webbing between the toes. In Central Australia the extent of toe webbing is variable but, at its extreme, is fleshy and reaches the tip of the toes. In *L. ornatus* toe webbing is consistently only basal. There is also a tendency for the brown markings upon the body to be more intense. In other respects *L. spenceri* conforms to the description of *L. ornatus*.

SIZE: Males 29–45 mm; females 33–46 mm.

HABITAT: Mostly associated with sandy watercourses and areas of inundation in sandy soils of the arid zones of northern and central Australia. In creek beds where the sand is still moist they emerge from their burrows at night, and the sites of re-entry can be detected as small, lipped depressions upon the surface.

ADVERTISEMENT CALL: Described as a rapid "ho-ho-ho-ho" with the o as in cot.

BREEDING AND LARVAE: Breeding takes place following heavy rain. Up to 1 130 eggs are laid in a foam nest. Tadpole development can be completed in 40 days and the tooth formula is $\frac{1\ ^{1}\ 1}{1\ _{2}\ 1}$.

SIMILAR SPECIES: As discussed above, the closest similar species is *L. ornatus*, but note that the tadpole of *L. spenceri* has two, instead of four, rows of upper labial teeth.

Limnodynastes tasmaniensis, Glen Lyon, Queensland *(Photo by Gordon Grigg)*.

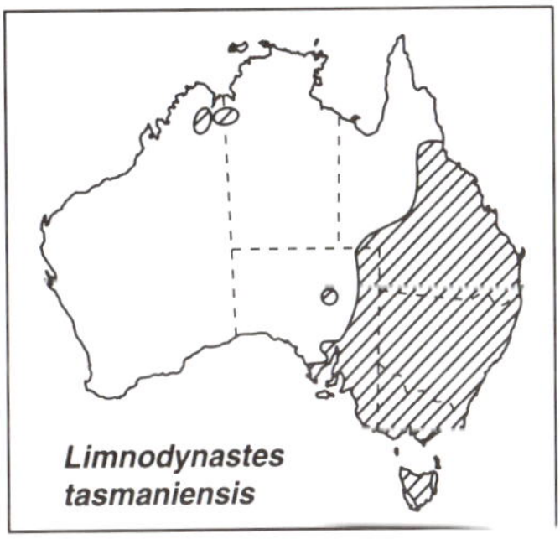

Limnodynastes tasmaniensis

Günther

Spotted Grass Frog

This is a very common species occurring over most of southeastern Australia and extending along the eastern seaboard. Its presence at Kununurra in northern Western Australia is believed to be the result of an accidental introduction. The dorsal surface has an attractive pattern of irregular olive-green or brown spots on a pale greyish-brown or dull green background. In many specimens there is a pale mid-dorsal stripe (red flanked by yellow in southeastern South Australia). The ventral surface is smooth and white except for males in breeding condition, when their throats are dusky yellow. The snout is moderately pointed with a white supralabial gland. The pupil is golden and the iris is subcircular. The tympanum is indistinct and, as in all *Limnodynastes*, the vomerine teeth are behind the choanae. The limbs are patterned similarly to the body and the toes are webbed slightly. There is a small inner metatarsal tubercle. The fingers lack webbing, with relative finger-length $3 > 2 > 4 > 1$. Females have distinct flanges on the fingers during the breeding season, whereas males have brown nuptial pads on the first and second fingers.

SIZE: Males 31–42 mm; female: 32–47 mm.

HABITAT: This species is widespread. It is found typically in marshy country, particularly in the vicinity of grass-lined streams and ponds, or flooded paddocks. By day *L. tasmaniensis* is found sheltering under logs, stones and debris, usually within a few metres of the water's edge.

ADVERTISEMENT CALL: There are three different call races, termed "northern", "southern" and "western". In the extreme south of New South Wales and in Victoria the call is a simple short, sharp "click", while elsewhere the call is a harsher series of groups of rapid clicks, thus "uk-uk-uk-uk-uk-uk" like a small boy imitating a machine gun. Just north of the Murray River an intermediate call-type can be recognized, consisting of two or three rapid "clicks". (Tape track 35.)

BREEDING AND LARVAE: Males call from the edge of shallow water, partly concealed by vegetation. They frequently make use of temporary waters such as flooded paddocks or roadside ditches resulting from heavy rains. Hundreds of calling males have been observed in these situations, their staccato cries filling the air and drowning conversation. Given appropriate rains, breeding can occur at any time of year, but the main breeding season falls between August and March. The foam nest of 90–1 350 eggs is 5–8 cm in diameter and floats in the water attached to emergent vegetation. After about three days, small black tadpoles hatch. These may be dark brown when they are older and reach about 46 mm in length. The larval life is 3–5 months and the tadpole tooth formula $\frac{4\ ^{1}\ 4}{1\ _{2}\ 1}$.

SIMILAR SPECIES: *Limnodynastes tasmaniensis* and *L. fletcheri* are often confused and the problem is aggravated by broadly overlapping distribution. Differences between the two species can be tabulated as follows:

	L. tasmaniensis	*L. fletcheri*
Rose-coloured eyelids	Absent	Usually present
Dorsal markings	Regular shape	Irregular shape
Dorsal stripe	Often present	Nearly always absent
Web on toes	Vestigial	Small, distinct
Finger lengths	$3 > 2 > 4 > 1$	$3 > 1 > 2 = 4$

Limnodynastes terraereginae

(Fry)

Northern Bullfrog

The dorsum is very dark brown or black, usually with a faint mid-dorsal stripe. The ventral surface is white or pale yellow and the groin yellow and scarlet. The gland below the eye is bright yellow. The toes are very slightly webbed. The tibial gland is prominent.

SIZE: Males 60–76 mm; females 55–79 mm.

HABITAT: The edge of dams, swamps and other areas of static water where there is adequate cover in the form of grass and other dense vegetation.

ADVERTISEMENT CALL: A very short "bonk" higher in pitch than the others of the *dorsalis* complex.

BREEDING AND LARVAE: Breeding commences about October and continues until May. The spawn and larvae are similar to *L. dorsalis* except that the tooth formula is $\frac{4\ ^{1}\ 4}{1\ _{2}\ 1}$.

SIMILAR SPECIES: *Limnodynastes terraereginae* resembles *L. interioris* more than other species in eastern Australia, but can be distinguished by having scarlet coloration in the groin and less webbing.

Limnodynastes terraereginae, Collinsville, Queensland *(Photo by Gordon Grigg) (Specimen courtesy Margaret Davies and Graeme Watson).*

Limnodynastes terraereginae, ventral surface, Grafton, New South Wales *(Photo by Gordon Grigg) (Specimen courtesy Graeme Watson).*

Genus *Megistolotis* Tyler, Martin and Davies

Megistolotis can be translated literally as "largest ears", alluding to the extremely large and prominent tympanum. It is also the only genus of frogs confined to the stony escarpment country of the Kimberley Division of Western Australia and Arnhem Land.

The genus contains only a single species. It is characterized by its exceptionally large tympanum and by having a male nuptial pad consisting of a cluster of large, black spines.

Megistolotis lignarius

Tyler, Martin and Davies

Woodworker Frog

The body is bulky with short, muscular limbs. The head is short and has a very large tympanum that is about the same diameter as the eye. There is a broad glandular fold from the posterior corner of the eye, passing across the upper margin of the tympanum and descending to a position anterior to the forearm. The fingers and toes are unwebbed and have prominent subarticular tubercles. The male nuptial pad consists of a clump of separate small spines capped in black. The dorsum ranges from dull slate to dark olive, occasionally densely mottled with darker or lighter pigment. The ventral surface is white or purplish-brown.

SIZE: Males 43–62 mm; females 47–61 mm.

HABITAT: We have found this species in only two situations. In the dry season it lives deep inside caves, whilst in the wet season individuals have been found beneath large, flattened boulders at the edge of rock-strewn creek beds, or on rock platforms at the foot of escarpments.

ADVERTISEMENT CALL: A single, short, soft tap resembling the sound produced when a piece of timber is struck with a hammer. This call is repeated at regular intervals and has given the species its common name.

Megistolotis lignarius, Lake Argyle, Western Australia *(Photo by Bohdar Stankowich-Janusch).*

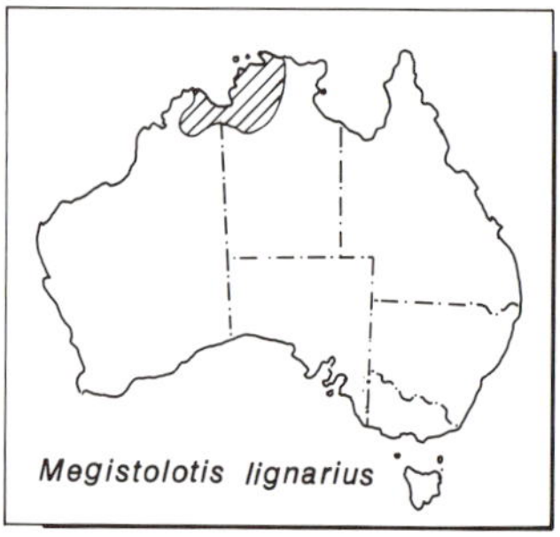

BREEDING AND LARVAE: Approximately 350 eggs are laid in a foam nest deposited in a rock pool. As the tadpoles grow they change from dark brown to intense black. The tadpoles complete their development in approximately nine weeks. The labial tooth formula is variable, being one of the following:

$$\frac{2\ {}^{2}\ 2}{3} \qquad \frac{3\ {}^{2}\ 3}{3} \qquad \frac{4\ {}^{2}\ 4}{3}$$

SIMILAR SPECIES: The very large tympanum, approximately equivalent to the horizontal diameter of the eye, is a feature unique to *M. lignarius* and is proportionately much larger than in any other Australian species.

Genus *Heleioporus* Gray

Heleioporus is a genus of moderate to large burrowing frogs distributed principally in the south-west of Western Australia (five species), with a single species in southeastern Australia. All members of this genus have stout bodies with broad rounded heads, large bulging eyes and short limbs without webbing between the toes. A distinguishing characteristic for most species of the genus (except *H. eyrei* and some *H. psammophilus*) is the presence of black conical nuptial spines on the first and occasionally second and third fingers of the male. The toes are short and stubby with only a trace of fleshy web. The inner metatarsal tubercle is greatly enlarged. The pupil is vertical and the tympanum is usually distinct. Vomerine teeth are in two transverse series between or slightly behind the choanae. The sternum is bifurcated posteriorly and this can be seen through the ventral skin.

In all species, males call from burrows and it is here that amplexus and oviposition occur. The spawn is a frothy mass containing up to 600 large eggs. Embryos develop in the froth until hatching, which may be delayed for a considerable period of time until flooding of the burrow occurs. The mating call of all species is broadly similar in structure, differing in frequency, call duration, and note and call repetition rate. Accurate identification of certain species is often very difficult using visible features only and differences in the mating call and geographic distribution are often the only means of identification. The key that follows relates only to male *Heleioporus*.

Key to male *Heleioporus*

1a. First finger with at least one black nuptial spine 2
b. First finger without black nuptial spine 6
2a. Dorsum and sides without distinct spotting or marbling .... *H. inornatus* (part) (p. 210)
b. Dorsum and/or sides with large spots or marbling 3
3a. Flanks with yellow spots. Dorsum uniform with or without tiny black spines capping warts 4
b. Other than above 5
4a. A divided flap in anterior corner of eye. Inner metatarsal tubercle less than half the length of fourth toe. Eastern Australia .... *H. australiacus* (p. 206)
b. A double row of papillae in anterior corner of eye. Inner metatarsal tubercle at least half the length of fourth toe. South-west of Western Australia ..... *H. barycragus* (p. 208)
5a. Dorsum with yellow or white spots on dark background .... *H. albopunctatus* (p. 204)
b. Dorsum marbled with brown to white on darker background .... *H. inornatus* (part) (p. 210)
6a. Mating call a long series of rapidly-repeated pulses "put-put-put" .... *H. psammophilus* (p. 212)
b. Mating call a long low moan .... *H. eyrei* (p. 208)

Females of *H. australiacus, H. barycragus, H. albopunctatus* and uniform-coloured *H. inornatus* can be distinguished by the final diagnostic character for each species in the key to males. It is nearly impossible to distinguish between female *H. eyrei, H. psammophilus* and marbled *H. inornatus*. The problem is compounded in that male *Heleioporus* lack vocal sacs.

Heleioporus albopunctatus

(Gray)

Western Spotted Frog

This is a large and distinctive species. The dorsum is dark chocolate brown to black with conspicuous yellow or white spots scattered over the back and sides. The ventral surface is white. There is a vertical white stripe at the tip of the snout and another from below the anterior margin of each eye to the angle of the jaw. The tympanum is distinct and vertically oval. Breeding males have one large and several small conical black spines on the first finger. The inner metatarsal tubercle is large and is at least one half of the length of the fourth toe.

Heleioporus albopunctatus, Walpole, Western Australia *(Photo by Gordon Grigg)* *(Specimen courtesy Grahame Webb).*

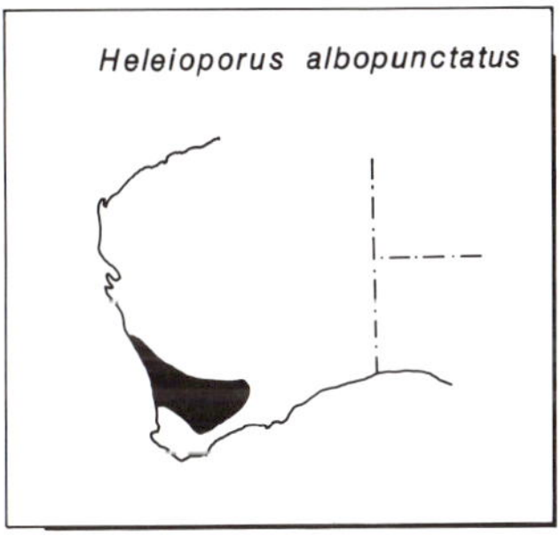

SIZE: Males 56–77 mm; females 60–85 mm.

HABITAT: This species is found only in the south-west of Western Australia. Adults are active on the surface during summer and autumn nights. Their burrows are usually found in the banks or beds of sandy watercourses.

ADVERTISEMENT CALL: A short high-pitched "coo-coo" repeated slowly.

BREEDING AND LARVAE: Calling begins with the advent of winter rains in April or May. The method of spawning has been described in the introduction to this genus and applies to all members of *Heleioporus*. 250–700 eggs are laid. The tadpole tooth formula is $\frac{5\ ^{1}\ 5}{1\ _{2}\ 1}$.

SIMILAR SPECIES: The yellow or white spots on a dark dorsal surface distinguish *H. albopunctatus* from others in the genus.

Heleioporus australiacus

(Shaw and Nodder)

Giant Burrowing Frog

The only member of the genus found in southeastern Australia, *H. australiacus* is a large species. The dorsal colour is dark chocolate brown to bluish-black. The skin is covered with small warts, each of which is capped with a tiny black spine. There are a number of yellow spots on the sides. The ventral surface is white or bluish-white. The head is rounded and very broad and there is a short yellow ridge beneath the eye and tympanum. The muscles of the forelimbs of breeding males are greatly enlarged and there are one large and several small black conical spines on the first finger. The second and third fingers also have small spines. The inner metatarsal tubercle is enlarged for burrowing.

SIZE: Males 60–78 mm; females 73–97 mm.

HABITAT: This species is most common on the Hawkesbury Sandstone in the Sydney region. It also occurs south to Victoria. Males begin calling during the first rains of August when temperatures are above 13°C and calling continues on damp nights until late March. Males call from burrows which are situated in sandy banks close to the water. They are usually associated with crayfish burrows.

ADVERTISEMENT CALL: An eerie, owl-like "oo-oo-oo".

BREEDING AND LARVAE: Except for the time of breeding, spawn and larvae are similar to *H. albopunctatus*.

SIMILAR SPECIES: *H. australiacus* and *H. barycragus* are very similar but their distinct distribution eliminates any chance of confusion. They can, however, be separated by the shape of the papillae in the anterior corner of the eye and the length of the inner metatarsal tubercle.

In Western Australia *Limnodynastes dorsalis* and in south-east Australia *L. dumerilii* may be confused with these species. *Heleioporus* is distinguished from *Limnodynastes* by the vertical pupil, distinct tympanum, and the vomerine teeth which do not extend beyond the outer margins of the choanae in *Heleioporus,* whilst *Heleioporus* lacks the conspicuous calf glands characteristic of these species.

'eleioporus australiacus, Kuring-gai Chase, New South Wales *(Photo by Gordon Grigg)*.

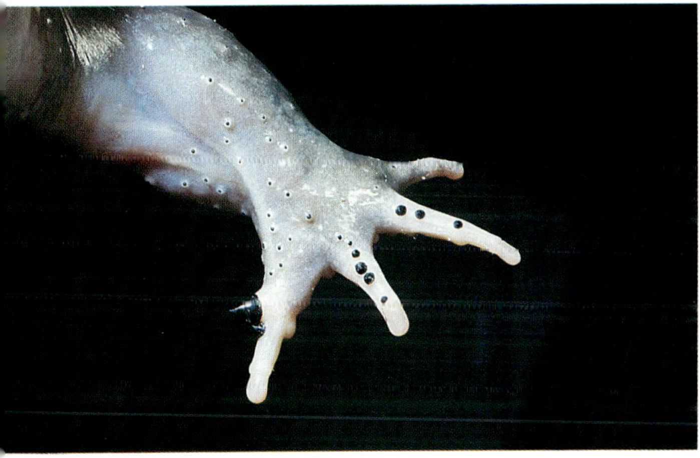

Ieleioporus australiacus, male nuptial spines, Kuring-gai Chase, New South Wales *(Photo by Gordon Grigg)*.

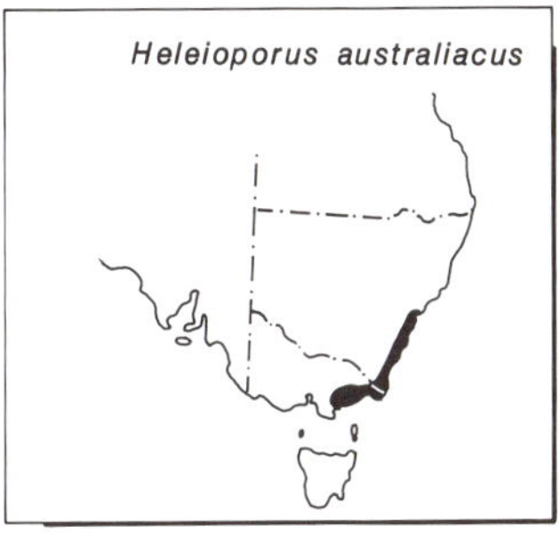

Heleioporus barycragus

Lee

Western Marsh Frog

For many years this species was not realized to be distinct from *H. australiacus,* despite their disjunct distribution on opposite sides of the continent. It differs in the lack of spines capping the dorsal warts, the shape of the papillae in the anterior corner of the eye (a large flap rather than distinct rows of papillae) and the length of the inner metatarsal tubercle which is at least one half of the length of the fourth toe.

SIZE: Males 60–83 mm; females 62–86 mm.

HABITAT: The burrows of this species are found in the banks of temporary watercourses on the slopes of the Darling Range.

ADVERTISEMENT CALL: Calls during autumn. A low-pitched, owl-like "hoot" repeated slowly.

BREEDING AND LARVAE: Similar to *H. albopunctatus* except that the tadpole tooth formula is $\frac{3\ ^{2}\ 3}{1\ _{2}\ 1}$.

SIMILAR SPECIES: See above and *H. australiacus.*

Heleioporus eyrei

(Gray)

Moaning Frog

This and the two following species are exceptionally difficult to tell apart except by the male mating call. The dorsum is brown with a lighter marbling of yellow, white, or grey. The ventral surface is white. There is a pale nodular ridge under the eye and a series of white spots on the sides. Limbs are not as robust as in the previous three species and males lack the nuptial spines. The inner metatarsal tubercle is enlarged.

SIZE: Males 45–66 mm; females 46–63 mm.

HABITAT: The burrows of this species are found in sandy swamps in the extreme southwestern corner of Western Australia. *H. eyrei* is found also on Rottnest and Bald Islands.

eleioporus barycragus, Darling Range, Western Australia *(Photo by Steve Donnellan).*

eleioporus eyrei, Walpole, Western Australia *(Photo by Gordon Grigg) (Specimen courtesy Grahame Webb).*

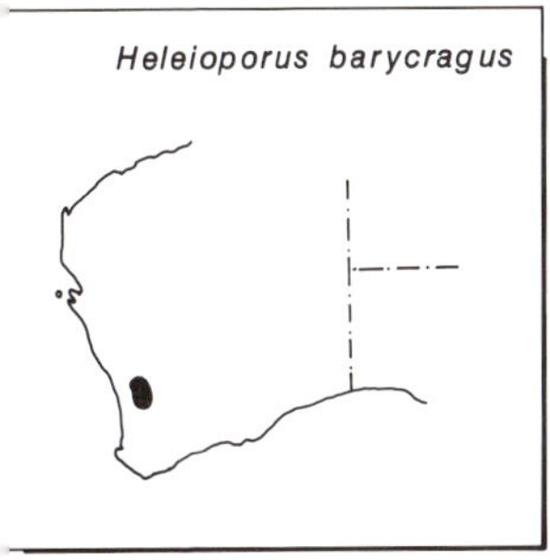

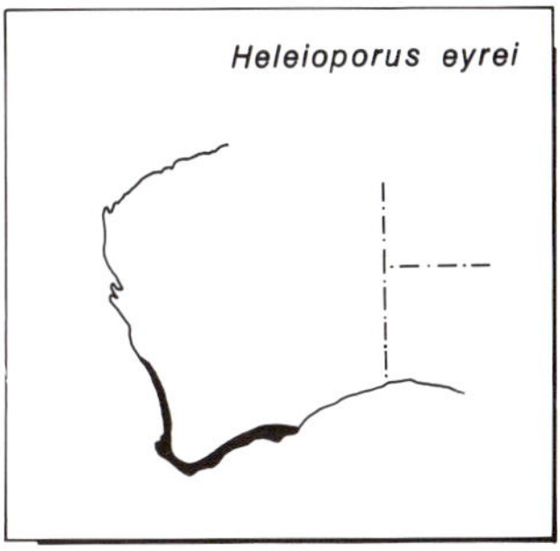

ADVERTISEMENT CALL: Calls during April and May. A long low moan repeated slowly.

BREEDING AND LARVAE: The egg mass includes 80–500 eggs. The tadpole tooth formula is $\frac{4\;^{1}\;4}{1\;_{2}\;1}$.

SIMILAR SPECIES: The three species, *H. eyrei*, *H. inornatus* and *H. psammophilus*, pose a problem of identification. Body length can be used as a guide because *H. eyrei* is generally larger than *H. psammophilus* and smaller than *H. inornatus*. The lack of breeding spines separates male *H. eyrei* from *H. inornatus* and the population of unspined male *H. psammophilus*. Most frequently *H. inornatus* has a uniformly dark dorsum. The greatest difficulty is to distinguish between *H. eyrei* and the unspined male *H. psammophilus*, the distinct differences in mating call providing the best guide.

Heleioporus inornatus

Lee and Main

Plains Frog

This is a moderately large frog. The most common colour phase is a uniform dark chocolate brown dorsum with tiny white spots on the lateral surfaces. A marbled form with pale markings on a brown background is found occasionally. Male *H. inornatus* have a few small breeding spines on the first finger. The tympanum is distinct and oval vertically.

SIZE: Males 44–64 mm; females 57–73 mm.

HABITAT: This species has a restricted distribution on the Darling Range and is usually found in sandy acid peat bogs where the principal vegetation is *Xanthorrhoea*.

ADVERTISEMENT CALL: "woop-woop-woop" during autumn.

BREEDING AND LARVAE: 100–250 eggs are laid. The tadpole tooth formula is similar to *H. eyrei*.

SIMILAR SPECIES: See *H. eyrei*.

eleioporus inornatus, Darling Range, Western Australia *(Photo by Mike Mahony).*

eleioporus psammophilus southwestern Western Australia *(Photo by Gordon Grigg) (Specimen courtesy Grahame Webb).*

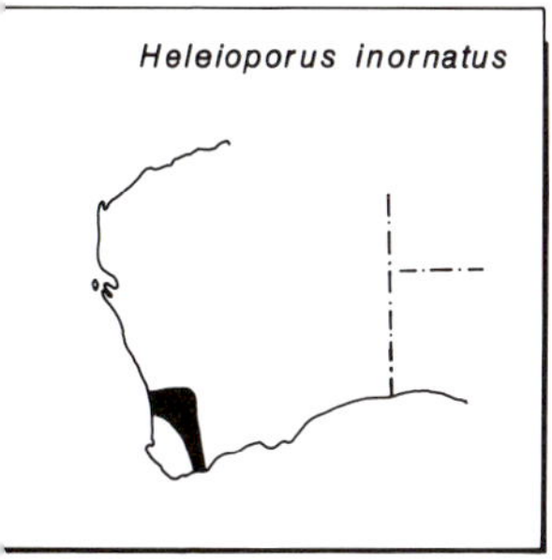

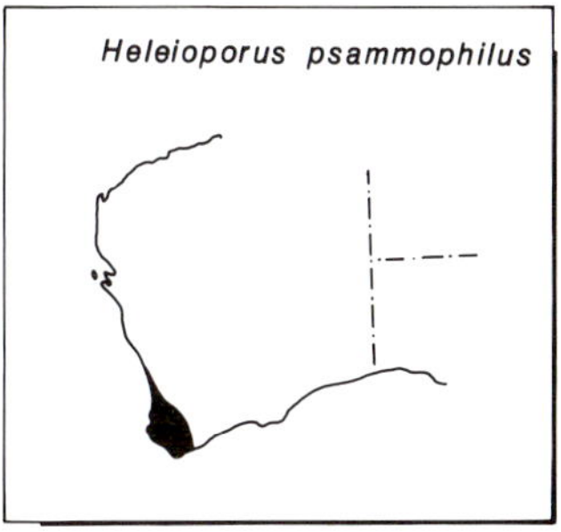

Heleioporus psammophilus

Lee and Main

Sand Frog

The dorsal surface is brown with white or grey marbling, the ventral surface white. There are numerous tiny white spots on the flanks and around the cloaca. Males have been found both with and without breeding spines. The tympanum is distinct and the inner metatarsal tubercle enlarged.

SIZE: Males 37–62 mm; females 52–61 mm.

HABITAT: Confined to sandy areas of the Darling Range and its foothills.

ADVERTISEMENT CALL: A short, high pitched "put-put-put-put . . " described as being like a small outboard engine. Calling is confined to April and May.

BREEDING AND LARVAE: Similar to *H. eyrei*.

SIMILAR SPECIES: See *H. eyrei*.

Genus *Neobatrachus* Peters

Identification of *Neobatrachus* species is often extremely difficult, though it has been simplified greatly by the work of Michael Mahony and Dale Roberts. As a result of their studies 10 species of *Neobatrachus* are recognized, four of which have been described since the publication of the first edition of this book. By far the greatest number of species is to be found in the south-western portion of Western Australia. The genus is absent from the extreme north of the continent, from most of Queensland and from Tasmania.

There is minimal variation in form: all species are squat, with large eyes and a vertical pupil. The limbs are short and the feet partially webbed, whilst the fingers are unwebbed.

Neobatrachus species burrow with the aid of a prominent, sharp inner metatarsal tubercle that is unpigmented in some species, edged with black or entirely black in others. Males lack a vocal sac and commonly have tiny black thorns upon the dorsal skin: presumably to reveal their sex to other males who grasp them by mistake.

Females lay approximately 1 000 eggs in chains in static water.

Key to *Neobatrachus* species in southeastern Australia

(South-east Queensland to the Eyre Peninsula in South Australia)

1a. Large species (46–52 mm), lacking a skin connection from the knee across the groin to the side of the body .... *N. pictus* (p. 222)

b. Smaller species (36–48 mm), with a distinct skin connection extending from the knee across the groin to the side of the body 2

2a. Dorsal surface rough with many small raised tubercles, greenish-yellow or brown with large dark brown patterning often with a fine yellow or red vertebral strip .... *N. sudelli* (p. 224)

b. Dorsal surface smooth with flattened tubercles, light brown or pale with darker irregular patterning .. *N. centralis* (p. 216)

Key to *Neobatrachus* species in western and central Australia

1a. Upper surface of foot white, inner metatarsal tubercle unpigmented or edged with light brown . *N. albipes* (p. 214)

b. Upper surface of foot mottled or transparent showing the underlying pink flesh colour, metatarsal tubercle pigmented with a black cutting edge 2

2a. Dorsal surface chocolate brown; on each side a pair of diverging, bright yellow stripes commencing behind the eye .... *N. wilsmorei* (p. 227)

b. Dorsal surface golden brown with darker marking, or a pattern of darker markings and reticulations in grey-brown, on a lighter brown or yellow background 3

3a. Distinct skin connection from knee across the groin to the side of the body 4

b. No distinct skin connection from knee across the groin .. 7

4a. Inner metatarsal tubercle pointed in profile; smaller species (34–42 mm), dorsal surface chocolate brown with lighter mustard yellow variegations; call a short trill .... *N. fulvus* (p. 218)

b. Inner metatarsal tubercle rounded in profile 5

5a. Smaller species (34–42 mm); dorsal surface smooth, golden or light brown overall with small dark spots or circles, often with a large butterfly-shaped marking in the post-scapular region. Call: a repetitive tapping .... *N. sutor* (p. 225)
b. Dorsal surface light brown or yellow background with a pattern of darker markings and reticulations. Call: a trill . 6

6a. Large species (48–59 mm); dorsal surface dark chocolate brown with mustard yellow mottling and a mid-vertebral stripe. Call: a short high-pitched trill . *N. aquilonius* (p. 216)
b. Moderate species (35–51 mm); dorsal surface light brown with irregular darker mottlings. Call: a short high-pitched trill .... *N. centralis* (p. 216)

7a. The edge of the metatarsal tubercle not strongly pigmented; moderate species (30–41 mm), dorsal surface greenish-yellow with grey marbling; fine red vertebral strip usually present .... *N. pelobatoides* (p. 222)
b. Metatarsal tubercle strongly edged in black; larger species (48–59 mm), dorsal surface yellowish to bronze with darker mottling .... *N. kunapalari* (p. 220)

Neobatrachus albipes

Roberts, Mahony, Kendrick and Majors

White-footed Frog

This is the most recently described member of the genus (1991). It is characterized by having a broad head and very large eyes, a visible tympanum and parotoid glands which can be detected as raised areas behind the head. The species name alludes to the white pigmentation on the upper surface of the feet, whilst the overall coloration ranges from grey to a light green-yellow.

SIZE: Males 33–45 mm; females 35 mm.

HABITAT: Known only from temporarily flooded areas in the wheatbelt of southwestern Australia.

ADVERTISEMENT CALL: A series of short, rapidly repeated pulses. Males call from concealed positions such as beneath bushes in flooded areas.

BREEDING AND LARVAE: Breeding follows heavy rain. Amplexus is inguinal but egg deposition and the tadpoles have not been observed.

Neobatrachus albipes, Ravensthorpe, Western Australia (Photo by Mike Mahony).

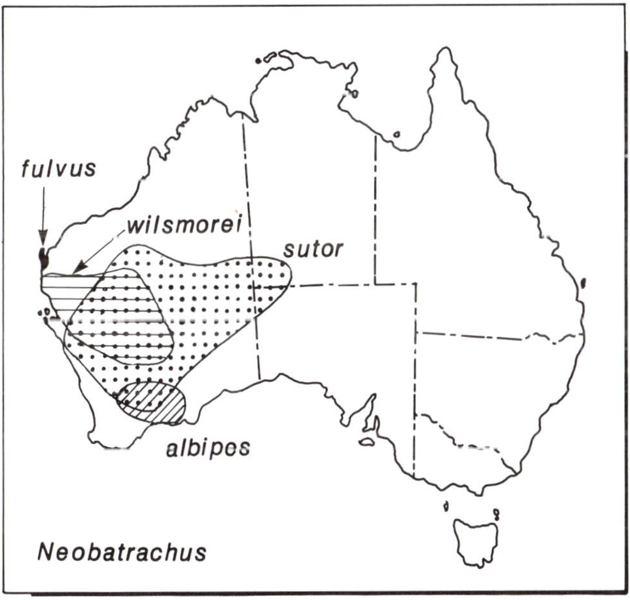

SIMILAR SPECIES: *Neobatrachus albipes* occurs with *N. pelobatoides* and *N. kunapalari*. From these species *N. albipes* is distinguished by having the upper surface of the foot white, and the inner metatarsal tubercle unpigmented or edged with lighter brown. *Neobatrachus albipes* has a shorter snout, larger tympanum and broader head than *N. pelobatoides*, and is smaller than *N. kunapalari*.

Neobatrachus aquilonius

Tyler, Davies and Martin

Northern Burrowing Frog

A robust species with a brilliant yellow or greenish background colour that is heavily suffused with black, dark brown or burgundy. The limbs are very short and plum coloured. The inner metatarsal tubercle is unpigmented. This species has the most northern distribution of all members of the genus.

SIZE: Males 48–54 mm; females 52–59 mm.

HABITAT: Grasslands and open forests.

ADVERTISEMENT CALL: A short, low trill.

BREEDING AND LARVAE: Breeding takes place in shallow flooded areas following summer rain. An individual laid 1 426 eggs. Tadpoles are unknown.

SIMILAR SPECIES: *Neobatrachus centralis* is the only species known to have a distribution approaching the southern limit of *N. aquilonius*. That species is distinguished by having a cream or grey background coloration, as opposed to yellow or greenish.

Neobatrachus centralis

(Parker)

Trilling Frog

At the periphery of its distribution *N. centralis* has been confused with a number of species. Hence, we consider that accounts in field guides and other publications are commonly inaccurate and should be treated with caution. The head is high and broad. The eye is large and the tympanum is not visible. The limbs are short

Neobatrachus aquilonius, Tanami Desert, Northern Territory *(Photo by Mike Mahony).*

Neobatrachus centralis, Strzeleki Track, South Australia *(Photo by Mike Mahony) (Specimen courtesy Harald Ehmann).*

and the fingers and toes are cylindrical. The hand is unwebbed and the fingers unfringed, whereas the toes are extensively webbed and bear lateral fringes. The overall coloration is sandy, grey or brown bearing irregular darker or lighter markings often with a pale mid-dorsal stripe.

SIZE: Males 41–50 mm; females 41–55 mm.

HABITAT: In arid regions associated with clay soils near sparse woodland and *Triodia* covered sandhills.

ADVERTISEMENT CALL: A loud, high-pitched trill.

BREEDING AND LARVAE: Tadpoles attain a maximum length of 57 mm. The tadpole tooth formula is $\frac{3\;^{1}\;3}{1\;_{2}\;1}$.

SIMILAR SPECIES: As indicated in the two distribution maps, *N. centralis* coexists only with *N. sutor* which is a much smaller species, has a bronze background colour and a tapping call. Literature reports of *N. centralis* from Western Australia are probably *N. kunapalari*, and Victorian records are *N. sudelli*.

Neobatrachus fulvus

Mahony and Roberts

Tawny Frog

The characteristic features of this species are short limbs, loose skin in the groin and a large inner metatarsal tubercle which is roughly rectangular in profile with rounded edges and a black upper margin. The head and body colour ranges from chocolate to dull claret, bearing small spots of irregular patches of mustard or pale yellow, most dense upon the flanks.

SIZE: Males 38–42 mm; females 42–48 mm.

HABITAT: Located on red or brown sandy loam, with breeding congresses in flooded claypans in the North West Cape region of Western Australia.

ADVERTISEMENT CALL: A short, high-pitched trill.

BREEDING AND LARVAE: Breeding has been observed in May following heavy rain. Details of the spawn and tadpoles are unknown.

Neobatrachus fulvus, Exmouth, Western Australia *(Photo by Mike Mahony)*.

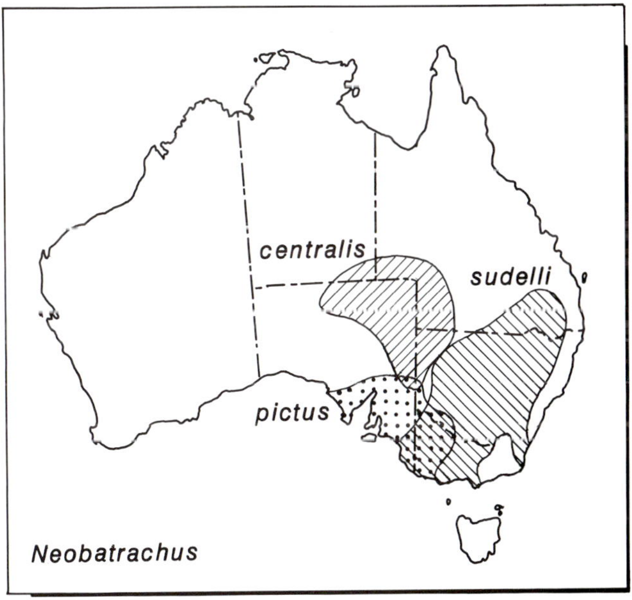

SIMILAR SPECIES: The distinctive coloration will distinguish it from other species in Western Australia. For example, *N. aquilonius*, which has the dorsal surface mottled with bright yellow and green; *N. sutor*, which is gold or pale brown with distinct darker spots; *N. wilsmorei*, which has a pair of diverging yellow stripes extending from the eye along the flanks; and *N. kunapalari*, which has a pale yellow to bronze background coloration, bearing isolated black patches.

Neobatrachus kunapalari

Mahony and Roberts

Kunapalari Frog

This is one of the largest species of *Neobatrachus*, reaching a maximum recorded length of 59 mm. The dorsal surface of males, particularly the anterior half, is covered with small, pointed tubercles that develop into spines during the breeding season. The inner metatarsal tubercle is large, rounded in profile and bears a sharp, black tip. The dorsal coloration ranges from yellow to bronze with clearly defined islands of black or brown.

SIZE: Males 48–58 mm; females 53–59 mm.

HABITAT: Common in areas of clay soils that are lightly vegetated.

ADVERTISEMENT CALL: A high-pitched trill of 18–21 pulses lasting for about 0.8 sec.

BREEDING AND LARVAE: Breeding occurs in summer or winter, and a total of 1 096 eggs were laid by one individual. Tadpoles are dull cream or grey and can reach a total length of 75 mm. The tadpole tooth formula is $\frac{2\ ^{1}\ 2}{1\ _{2}\ 1}$.

SIMILAR SPECIES: The coloration of *N. kunapalari* distinguishes it from *N. aquilonius* which is brightly marked with green and yellow. Similarly, the presence of paired diverging yellow stripes along the side of the body of *N. wilsmorei* distinguishes that species. *Neobatrachus sutor* shares with it the feature of having islands of dark pigment upon a paler background but with a size range of 35–51 mm, is smaller.

eobatrachus kunapalari, Ninghan, Western Australia *(Photo by Steve Donnellan)*.

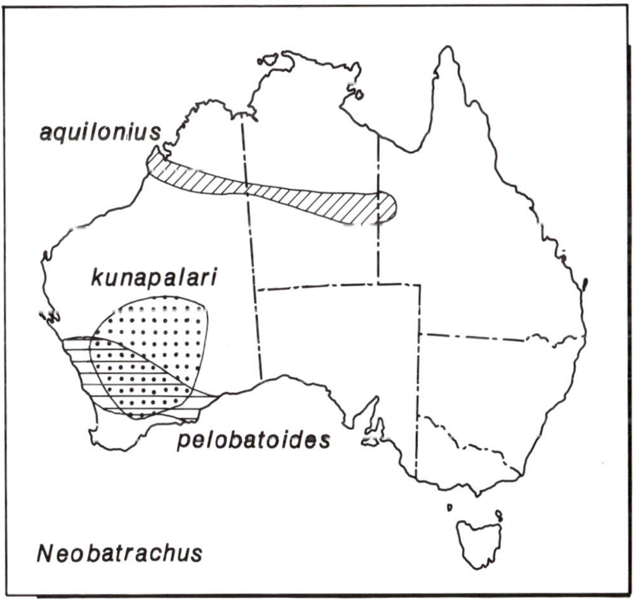

Neobatrachus pelobatoides

(Werner)

Humming Frog

The background colour ranges from pale yellow, with faint buff patches, to pale brown with sharply demarcated islands of olive. In some species a fine red or white stripe extends from the tip of the snout to the vent. The ventral surface is white. The toes are almost one-third webbed except in breeding males where the web extends to the tip on a tapering fringe. The inner metatarsal tubercle is edged in brown or black.

SIZE: Males 37–43 mm; females 36–44 mm.

HABITAT: Associated with clay or loamy soils.

ADVERTISEMENT CALL: A long, low-pitched trill or hum.

BREEDING AND LARVAE: Breeding occurs in autumn or early winter. Approximately 120 days are spent in the tadpole stage.

SIMILAR STAGES: This rather nondescript species is identified more readily by the lack of the unique features that distinguish other members of the genus. The lack of the diverging lines distinguish it from *N. wilsmorei* and the nature of the coloration from other species within the genus.

Neobatrachus pictus

Peters

Painted Frog

This is a stockily built large species, generally deep olive on the head and body, with diffuse darker markings. The eye is particularly prominent and has a vertical pupil. The tympanum is not visible. The limbs are short. The fingers are cylindrical and lack fringes and webbing, whereas the toes have broad flanges and extensive webbing between them. The inner metatarsal tubercle is black. The dorsal skin is smooth, except during the mating season when the skin of the male develops tiny, scattered, black thorns, whilst the nuptial pad extends over the first two or even three fingers.

eobatrachus pelobatoides, Quairading, Western Australia *(Photo by Mike Mahony).*

'eobatrachus pictus, Robe, South Australia *(Photo by Mike Mahony).*

SIZE: Males 46–58 mm; females 48–55 mm.

HABITAT: Open grassland, woodland.

ADVERTISEMENT CALL: A long, musical trill.

BREEDING AND LARVAE: Breeding takes place in autumn or winter. Males call whilst floating in static water. Approximately 1 000 eggs are laid in a chain entwined in submerged vegetation. The tadpole tooth formula is $\frac{2\,^{1}\,2}{1\,_{2}\,1}$.

SIMILAR SPECIES: *Neobatrachus sudelli* is encountered within a portion of the geographic range of *N. pictus*. It is a smaller species, has a shorter call and looser skin in the groin. *Neobatrachus pictus* is distinguished from *N. centralis* by its olive coloration and entirely black inner metatarsal tubercle (at most edged with black in *N. centralis*).

Neobatrachus sudelli

(Lamb)

Sudell's Frog

This species is particularly distinctive amongst those of the south-east of the continent because of the clearly demarcated pattern of markings upon the dorsal surface. These markings may be olive or pale green on a background of dark brown or tan with a pale mid-dorsal stripe. There is a membrane of skin between the knee and the side of the body.

SIZE: Males 38–44 mm; females 46–49 mm.

HABITAT: Open grassland and lightly wooded country.

ADVERTISEMENT CALL: A short musical trill.

BREEDING AND LARVAE: The male calls whilst floating in shallow, still water. The spawn is in the form of elongate strands that become entangled in aquatic vegetation. The tadpole is pale grey with a metallic sheen. A span of 4½–7 months has been recorded for tadpoles. The tadpole tooth formula is $\frac{2\,^{1}\,2}{1\,_{2}\,1}$.

eobatrachus sudelli, Inverell, New South Wales *(Photo by Gordon Grigg)*.

SIMILAR SPECIES: The dorsal markings distinguish this species from the larger *N. pictus*. The pattern is such it could be mistaken for *Limnodynastes tasmaniensis*, from which it can be separated by the form of the inner metatarsal tubercle.

Neobatrachus sutor

Main

Shoemaker Frog

The background colour of this species is gold, yellow or pale bronze on which are numerous, small, discrete islands of black or dark brown. The inner metatarsal tubercle has a black outer margin. The toes are fully webbed.

SIZE: Males 35–42 mm; females 34–51 mm.

HABITAT: Areas of clay or loamy soils.

ADVERTISEMENT CALL: A short series of taps, giving rise to the name (*sutor*: shoemaker).

BREEDING AND LARVAE: Breeds in claypans following summer rains. Forty days are spent in the tadpole stage.

Neobatrachus sutor, Shark Bay, Western Australia *(Photo by Mike Mahony)*.

Neobatrachus wilsmorei, north of Geraldton, Western Australia *(Photo by M[illegible] Mahony)*.

SIMILAR SPECIES: Several species of *Neobatrachus* have a back pattern of small islands of dark pigment upon a paler background. Of these *N. sutor* is the smallest and it has a characteristic call.

Neobatrachus wilsmorei

Parker

Goldfields Bullfrog

This distinctive species is characterized by having fine, longitudinal yellow lines along the back, one narrow mid-ventral stripe and, on each flank, a pair of lines that diverge posteriorly, one branch passing obliquely to the cloacal region and the other continuing along the flank to the groin. The background colour is various shades of brown. The toes are about two-thirds webbed with an enlarged inner metatarsal tubercle edged in black. Breeding males have a nuptial pad on the first finger.

SIZE: Males 56–61 mm; females 55–63 mm.

HABITAT: Claypans.

ADVERTISEMENT CALL: Described as "plonk-plonk-plonk".

BREEDING AND LARVAE: Breeds in flooded claypans following heavy summer rains. It takes about 40 days to complete development. The tadpole reaches a maximum length of 64 mm.

SIMILAR SPECIES: The fine longitudinal lines along the body are a feature not shared by any other species.

Genus *Mixophyes* Günther

This genus contains the largest myobatrachid frogs found in Australia. Their haunts are exclusively the rainforest and wet sclerophyll forest of the eastern coast and tablelands. Four species overlap broadly in southeastern Queensland and New South Wales: *M. fasciolatus, M. iteratus, M. balbus* and *M. fleayi.* A fifth species, *M. schevilli,* is restricted to north-eastern Queensland, which, according to Richards (1993) is a complex of three species, whilst another species is found in New Guinea. Most species are commonly known as Great Barred Frogs. There is concern regarding the decline in numbers of some species in south-east Queensland and north-east New South Wales.

The genus *Mixophyes* is distinguished by a vertical pupil; ear fully developed with a distinct tympanum; maxillary and vomerine teeth present; toes webbed, the web penetrating between the outer metatarsals; fingers free.

Key to *Mixophyes* species

1a. Toes webbed to tip of third toe on outer margin. Two joints of fourth toe free. Length of inner metatarsal tubercle half that of its distance from the tip of the first toe 2
b. Three joints of fourth toe free from web. Inner metatarsal tubercle as long as its distance from tip of first toe 3

2a. Dark cross-bars on hind limbs broad with intervening lighter areas which join together on posterior surface of thigh to form uniform dark background with scattered pale spots .... *M. iteratus* (p. 234)
b. Dark cross-bars narrower than intervening light area, joining together on posterior of thigh to form diffuse speckled horizontal band .... *M. schevilli* (p. 234)

3a. Dark cross-bars on hind limbs narrow and distinct, broadening on anterior and posterior margins to form triangular patches visible from beneath. Flanks with dark spots .... *M. fasciolatus* (p. 229)
b. Dark cross-bars on hind limbs narrow and indistinct. No dark lateral spots on flanks .... *M. balbus* (p. 229)

4a. Ventral surfaces yellow .... *M. fleayi* (p. 232)
b. Ventral surfaces white .... *M. fasciolatus* (p. 229)

Mixophyes balbus

Straughan

Stuttering Frog

This frog grows to 80 mm. The dorsal surface is brown to olive-green and finely granular. A dark irregular mid-dorsal band extends to the rear from between the eyes. The band may be broken into irregular blotches. The brown coloration of the dorsal surface diffuses laterally to merge with the white ventral coloration. There is a conspicuous dark head-stripe beginning as a triangular patch in front of the nostril and continuing as a thin stripe curving over the tympanum. The tympanum is distinct, large, and slightly oval. The pupil is vertical and the iris golden above and dark below. In life, there is a pale blue crescent under the upper eyelid. Vomerine teeth are between or slightly in front of the choanae. The bars on the hind limbs are thin and indistinct. Toes are three-quarters webbed and the fingers free from web.

SIZE: Males 60–63; females 74–80 mm.

HABITAT. This frog occurs in rainforest in the mountains of New South Wales, southeastern Queensland and Victoria.

ADVERTISEMENT CALL: A short, soft grating trill of about 10 pulses. (Tape track 19).

BREEDING AND LARVAE: Similar to *M. fasciolatus*. The tadpole tooth formula is $\frac{10\ ^{1}\ 10}{1\ _{2}\ 1}$.

SIMILAR SPECIES: See p. 236.

Mixophyes fasciolatus

Günther

Great Barred Frog

This is a large long-legged frog with a very broad head. The dorsal surface is smooth or slightly granular, the colour a beautiful copper or tan. There is a dark mid-dorsal band beginning as a T or Y between the eyes and extending to the rear as an irregularly-shaped band. In some specimens the band may be broken into

Mixophyes balbus, Gosford, New South Wales *(Photo by Gordon Grigg).*

Mixophyes fasciolatus, Gold Creek, Queensland *(Photo by Gordon Grigg).*

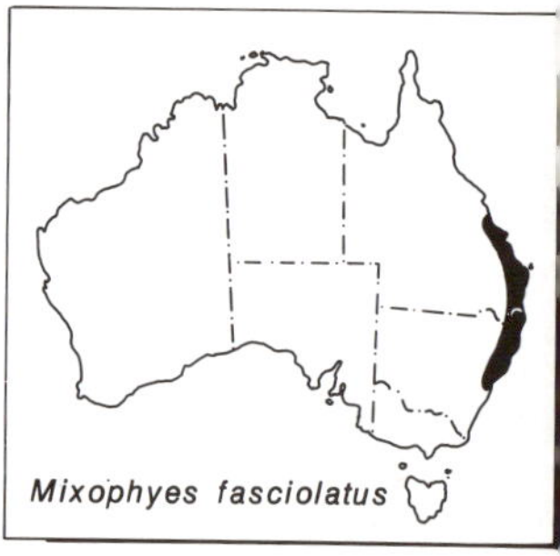

dark irregular blotches. On the sides, between the arm and groin, there is a series of dark irregular spots where the dark dorsal surface meets the light ventral surface. A conspicuous black head-stripe begins as a triangular patch in front of the nostril and continues as a thin stripe curving over the tympanum. The tympanum is distinct, large, and slightly oval. The pupil is vertical when constricted and the iris very dark brown. The vomerine teeth are between, or slightly in front of, the choanae. The legs are long and powerful, with characteristic bars which are thin and well defined, widening into triangular patches on the margins. The toes are three-quarters webbed. The amount of web is very important for positive identification (see diagram). The inner metatarsal tubercle is long and thin, equal in length to the first toe. The fingers are free from web.

SIZE: Males 60–65 mm; females 72–101 mm.

HABITAT: This species seems to prefer mountain or coastal rainforest and is often found on the sandy banks of mountain streams or among leaf litter.

ADVERTISEMENT CALL: A deep harsh "wahk". (Tape track 18).

BREEDING AND LARVAE: In late spring, eggs deposited among leaf litter on the banks of streams are washed into the water during heavy rain. The tadpoles are large (65 mm), with flattened bodies and narrow fins. The cross-bars are quite easy to see once the hind limbs are developed. The horny jaws contain three upper, three lower, and nine short, divided lateral rows. The mouth is surrounded by a broad band of papillae indented at the angle of the jaws.

SIMILAR SPECIES: See p. 236.

Mixophyes fleayi

Corben and Ingram

Fleay's Barred Frog

The dorsal surface is light brown and bears faint, irregular marbling of darker pigment. Between the eyeballs is a Y-shaped marking extending down the middle of the back to the groin. The sides of the body are greyish-brown fading inferiorly to cream and partially obscured by numerous, small, black spots. There is an irregular, continuous black patch from the nostril to above the insertion of the arms, and a dark purple patch beneath the eye. The undersurfaces of the body and limbs are cream-yellow. The soles and palms are black, the thighs bear 7–8 narrow, black cross-bands and the backs of the thighs are grey-brown. The pupil is vertical. The upper half of the iris is pale blue or occasionally silver. The toes are webbed to the base of the terminal phalanges of the first, second and fifth, while two phalanges of the third and those of the fourth are free from web.

SIZE: Males 63–70 mm; females 79–89 mm.

HABITAT: Confined to montane rainforest extending from the Conondale Range in south-east Queensland to the Richmond Range in north-east New South Wales.

ADVERTISEMENT CALL: Two distinct calls are made by this species. One is described as a throaty "ok-ok-ok-ok-ok-ok", and is the call made by solitary males from under leaf litter. However, in a chorus at the edge of water the call has been likened to a long, rasping "arrrrrrr".

BREEDING AND LARVAE: Unknown.

SIMILAR SPECIES: See p. 236.

Mixophyes fleayi, Cunninghams Gap, Queensland (*Photo by Owen Kelly*).

Mixophyes iteratus, Ourimbah, New South Wales (*Photo by Gordon Grigg*).

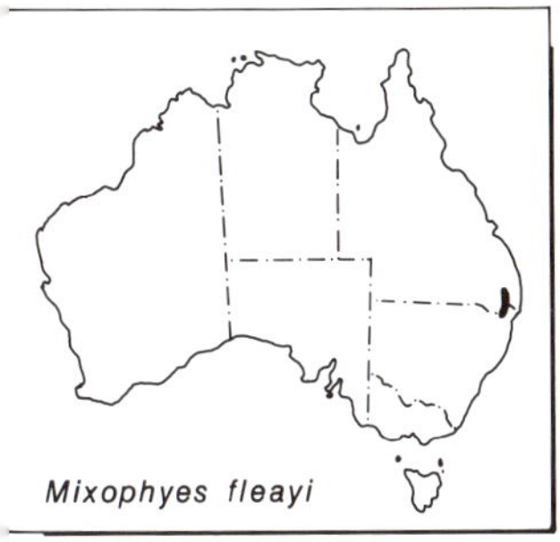

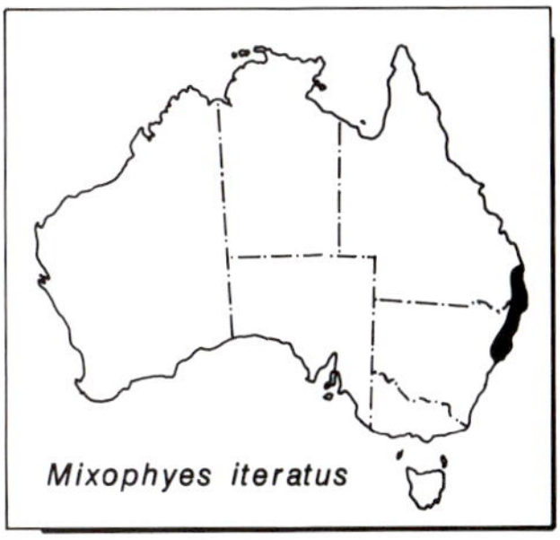

Mixophyes iteratus

Straughan

Southern Barred Frog, Great Barred Frog

The back is finely granular, the colour ranging from dark olive to black with the typical dorsal pattern (described in the key). The ventral surface is smooth and white except on the chin where there is fine mottling. The sides are irregularly spotted and mottled. The head is broad with the snout more pointed than in other *Mixophyes* species. The head-stripe is broad and almost the same width throughout, except over the tympanum. The tympanum is distinct, large, and circular. The colour of the iris is golden above and dark below. Vomerine teeth are between or slightly in front of the choanae. The light and dark bars on the hind limbs are of equal width and toes are fully webbed. The fingers, as in all species of this genus, are free from webbing.

SIZE: Males 68–78 mm; females 91–108 mm.

HABITAT: *Mixophyes iteratus* is found in the coastal riverine rainforest and upland areas such as the Border Ranges and Conondale Range.

ADVERTISEMENT CALL: A deep guttural grunt.

BREEDING AND LARVAE: Similar to *M. fasciolatus* but with only three or four lateral rows of teeth and a ventral gap in the band of papillae.

SIMILAR SPECIES: See p. 236.

Mixophyes schevilli

Loveridge

Northern Barred Frog

This species is closely related to *M. fasciolatus*. Maximum body length is 100 mm. The dorsum is copper or pale chocolate brown with a series of large dark irregular blotches along the middle of the back. The ventral surface is white. A dark stripe runs from the snout to the shoulder, curving over the tympanum. The powerful hind limbs have narrow distinct bands. Toes are nearly fully webbed, and the

Mixophyes schevilli, Tinaroo Creek Road, north Queensland *(Photo by Gordon Grigg).*

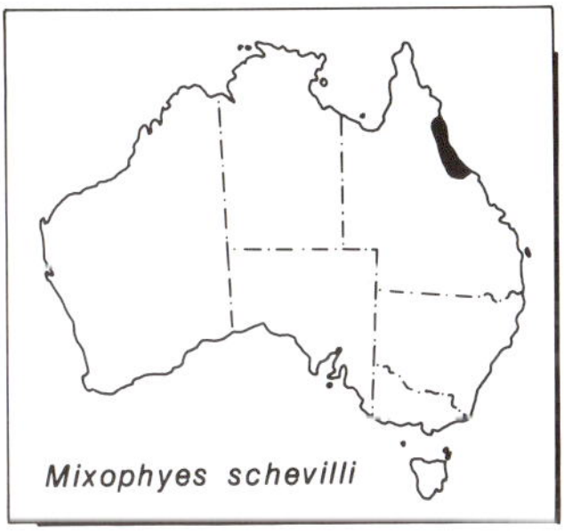

fingers are free. Vomerine teeth are between or slightly in front of the choanae. The inner metatarsal tubercle is about half the length of the first toe.

SIZE: Males 64–75 mm; females 71–87 mm.

HABITAT: *Mixophyes schevilli* lives around fast-flowing streams and nearby pools in mountainous rainforest in the Atherton Tableland district of northeastern Queensland.

ADVERTISEMENT CALL: A deep "wahk".

BREEDING AND LARVAE: The very large tadpoles can be found in most mountain streams of the area. Their total length reaches 125 mm. They have only four lateral rows of teeth.

SIMILAR SPECIES: See p. 236.

Identification of *Mixophyes* species

Any species of *Mixophyes* could perhaps be confused with either *Lechriodus fletcheri* or *Cyclorana novaehollandiae*. *Mixophyes* can be distinguished from *L. fletcheri* by its large size (80–115 mm), compared with 54 mm maximum, much greater webbing between the toes, absence of skin ridges along the body, and the position of vomerine teeth which occur behind the choanae in *L. fletcheri*. *Mixophyes* and *C. novaehollandiae* can easily be separated. The latter has a horizontal pupil, shorter legs, and reduced webbing on the toes. Differences between the five species of *Mixophyes* are shown in the following table:

	M. schevilli	*M. fasciolatus*	*M iteratus*	*M. balbus*	*M. fleayi*
Max. length	100 mm	80 mm	115 mm	80 mm	90 mm
Iris	Dark brown	Very dark brown	Upper half light (golden) Lower half dark	Same as *M. iteratus* but with sky-blue crescent under upper lid	As for M. *balbus*

Additionally, pay attention to the pattern of barring on the hind limbs and the extent of foot webbing (see individual species descriptions).

Genus *Notaden* Günther

Each of the four species of *Notaden* lives in a discrete portion of central or northern Australia and, as far as is known, no two species co-exist.The body is globular, the head short and the eyes prominent with a horizontal pupil. Webbing between the toes is slight or lacking. The inner metatarsal tubercle is spade-like. The skin is slightly warty and highly glandular, releasing a creamy adhesive when handled.

Key to *Notaden* species

1a. Inner metatarsal tubercle black ... *N. melanoscaphus* (p. 239)
 b. Inner metatarsal tubercle not black 2
2a. Back bearing striking yellow black and red markings .... *N. bennetti* (p. 237)
 b. Back not bearing yellow markings 3
3a. Back dull tan or brown and bearing scattered black tubercles .... *N. nichollsi* (p. 241)
 b. Back lacking black tubercles .... *N. weigeli* (p. 241)

Notaden bennetti

Günther

Holy Cross Toad

The most distinctive and brightly marked species of the genus, *Notaden bennetti* is rather like a colourful ping-pong ball. The snout is very short and the eyes widely spaced. The legs are short and stout but despite this *N. bennetti* can hop surprisingly rapidly. It is a very competent burrower. The dorsal background colour is bright yellow or greenish-yellow overlaid by a series of red, black and green warty protuberances forming a rough cross-shaped pattern on the back. The ventral surface and the limbs are white, the limbs often having a few black warts on the upper surfaces. The tympanum is hidden. Fingers and toes are short and stubby; the toes have a slight web.

SIZE: Males 42–63 mm; females 46–68 mm.

HABITAT: *Notaden bennetti* inhabits the slopes and plains to the west of the Great Dividing Range. Most of their lives are spent deep underground and they surface to breed only after heavy rain. Recent observations on stomach contents agree with early reports that the diet consists almost exclusively of ants and termites which they collect upon the surface.

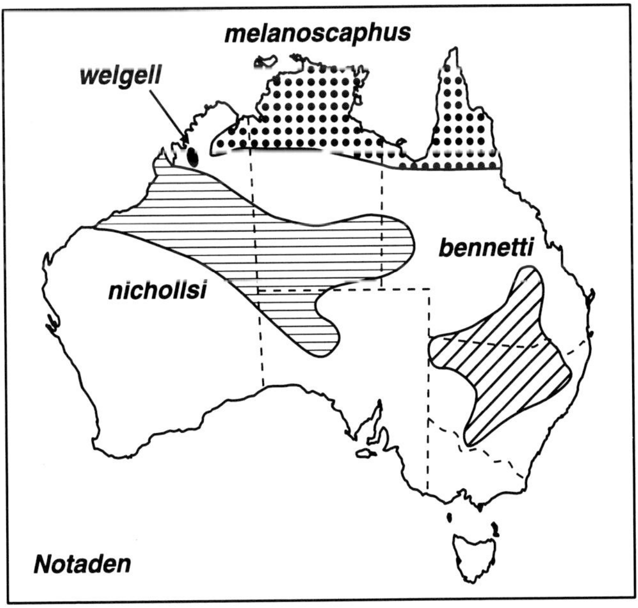

Notaden bennetti, Condoblin, New South Wales *(Photo by Gordon Grigg).*

Notaden bennetti, rear view to show cross, Condoblin, New South Wales *(Photo by Gordon Grigg).*

ADVERTISEMENT CALL: Males call while floating spreadeagled in water. The call is an owl-like "woop". (Tape track 45.)

BREEDING AND LARVAE: Little is known about larval development. The small pigmented eggs are deposited in water. Development is rapid, for most standing water is only temporary in the areas they inhabit.

SIMILAR SPECIES: *Notaden nichollsi* is the nearest relative of *N. bennetti*. The latter species is distinguished by the more distinct dorsal pattern and the short inner metatarsal tubercle.

Notaden melanoscaphus

Hosmer

Northern Spadefoot Toad

The markings on the back of this frog are often difficult to detect. When the background coloration is pale, a pattern of large, narrowly separated dark patches is revealed. This colour form is typical of juveniles. Generally the dorsal colour is a dull grey-brown. Unique to this species is the black pigmentation of the outer metatarsal tubercle. When disturbed this globular frog runs and, at a distance may be mistaken for a small rodent.

SIZE: Males 34–48 mm; females 45–49 mm.

HABITAT: Most commonly associated with open grassland in areas of clayey soils.

ADVERTISEMENT CALL: A loud continuous "whoop-whoop-whoop . . ." at a rate of one per second.

BREEDING AND LARVAE: Males call from shallow water, usually floating spreadeagled with their entire bodies inflated by their distended lungs. From 500 to 1 200 eggs are laid in long chains tangled in submerged vegetation. The tadpoles are dark brown and complete their development in eight weeks. The tadpole tooth formula is $\frac{1\ ^{1}\ 1}{1\ _{2}\ 1}$.

SIMILAR SPECIES: Geographically close to *N. weigeli* and *N. nichollsi* but distinguished by the black metatarsal tubercle.

Notadan melanoscaphus, Lake Argyle, Western Australia *(Photo by Bohdan Stankowich-Janusch).*

Notaden nichollsi, Port Hedland, Western Australia *(Photo by Mike Mahony).*

Notaden nichollsi

Parker

Desert Spadefoot Toad

The dorsum is olive-grey to brown with a number of yellow and red warts arranged in roughly linear rows. The ventral surface is pale. The limbs are short. Fingers are free from web; toes have a slight web. The tympanum is indistinct.

SIZE: Males 42–58 mm; females 46–65 mm.

HABITAT: Found in the dry north-west of Western Australia, the southern half of the Northern Territory, south-west Queensland and possibly the northern portion of South Australia, *N. nichollsi* has been dug from burrows at a depth of more than one metre.

ADVERTISEMENT CALL: A loud whooping sound.

BREEDING AND LARVAE: Eggs are laid in chains, enmeshed amongst submerged vegetation in temporarily flooded areas. The tadpoles are pale brown and take 16 days to complete development. The tadpole tooth formula is $\frac{2\ ^{1}\ 2}{1\ _{2}\ 1}$.

SIMILAR SPECIES: See *N. bennetti*.

Notaden weigeli

Shea and Johnston

Weigel's Toad

This is the largest and geographically most restricted species, being confined to a portion of the Kimberley in northern Western Australia. It was described in 1987 and nothing is yet known of its biology. It is a brightly coloured frog with a reddish-brown dorsum. There are numerous orange tubercles with white tips and other indistinct dark mottling. The ventral surface is grey.

SIZE: No male specimens are known, females 69–71 mm.

Notaden weigeli, Mitchell Falls, Western Australia *(Photo by John Weigel).*

HABITAT: This species has been found only upon rock platforms near the Mitchell River in northern Western Australia and at the entrance to a small gorge at Mount Elizabeth Station.

ADVERTISEMENT CALL: Unknown.

BREEDING AND LARVAE: Unknown.

SIMILAR SPECIES: The only *Notaden* species living near the range of *N. weigeli* is *N. melanoscaphus*, from which it can be distinguished readily by its pale (not black) outer metatarsal tubercle.

Genus *Philoria* Spencer

The single species of *Philoria* is confined to the slopes of Mt Baw Baw in Victoria. The restricted nature of its distribution, combined with habitat destruction associated with recreational facilities led to the species being declared endangered.

The relationship between *Philoria* and *Kyarranus* is very close, and recognition of the latter as a distinct genus has been questioned or even rejected (see Cogger 1992). However, there are osteological features (particularly of the pelvis) that separate them, and both genera have been found in fossil sites in north-west Queensland aged 20–30 million years. *Philoria frosti* is larger than any of the *Kyarranus* species and is also distinguished by having large parotoid glands.

Philoria frosti

Spencer

Baw Baw Frog

The head and body are flattened, the limbs short. The fingers and toes are unwebbed. The skin on the back bears numerous flattened warts, and there are paired, protuberant parotoid glands behind the head. The dorsal coloration ranges from dull brown to slate with poorly defined suffusions of pale yellow.

SIZE: Males 43–46 mm; females 47–55 mm.

HABITAT: Confined to the Mt Baw Baw plateau above 1 160 m, where it occurs on wet sub-alpine heath and montane riparian thickets.

ADVERTISEMENT CALL: A series of moans and grunts. Calling takes place between late October to late December.

BREEDING AND LARVAE: From 50 to 180 unpigmented eggs are laid in a foam nest concealed beneath a log or a large rock. The foam nest breaks down into a watery jelly and the hatching tadpoles remain within it. The tadpoles do not feed and lack tooth rows, living entirely upon their yolk reserves. The larval span occupies approximately 5–8 weeks.

SIMILAR SPECIES: There are no other species in the area that resemble *P. frosti* sufficiently to be confused with it.

ıiloria frosti, Mt Baw Baw, Victoria *(Photo by Gordon Grigg)*.

Philoria frosti, foam egg mass under rock, Mt Baw Baw, Victoria *(Photo Gordon Grigg)*.

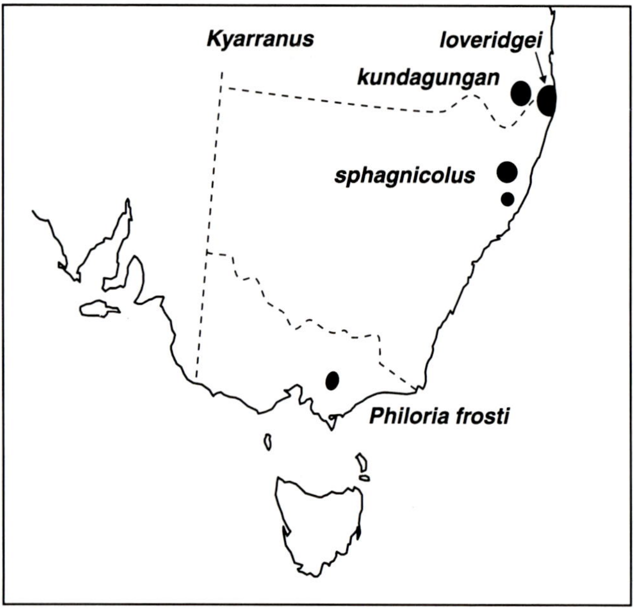

Genus *Kyarranus* Moore

This genus is part of a distinctive anuran fauna unique to montane moist forest areas of northeastern New South Wales and southeastern Queensland, although a fossil record from Riversleigh Station in the extreme north-west of Queensland indicates the enormous climatic changes that have occurred and that it was once distributed far more widely. The closest relative of *Kyarranus* is *Philoria* in eastern Victoria, so there may have been a common ancestor in the intermediate area. The three living species of *Kyarranus* are small to moderately sized secretive frogs living in areas of sphagnum and other water-saturated sites. They lay unpigmented eggs in a foam nest hidden from the light. The emerging tadpoles remain within the nest and live entirely upon their yolk.

Key to *Kyarranus* species

1a. Broad dark stripe on side of head 2
  b. No stripe on side of head .... *K. kundagungan* (p. 245)

2a. Brown band on flank, horizontal black band across cloacal region, throat heavily mottled ...... *K. sphagnicolus* (p. 249)
  b. No brown band on flank, no horizontal black band across cloacal region, throat lightly mottled . *K. loveridgei* (p. 247)

Kyarranus kundagungan

Ingram and Corben

Mountain Frog

Kyarranus kundagungan is a beautiful little rainforest frog which remained unknown until 1975. It is a small, robust and pear-shaped species. The dorsal colour is usually bright red but much darker specimens, almost black, have been found. There may be a black V-shaped mark on the back. The ventral surface is immaculate yellow, often with a small red patch on the throat. There is a black head-stripe from the naris, through the eye to the shoulder. The tympanum is indistinct and the fingers and toes are without webbing. The vomerine teeth are in two straight series posterior to the choanae.

Kyarranus kundagungan, Gambubal State Forest, Queensland *(Photo by Gordo Grigg).*

Kyarranus kundagungan, ventral view, Cunninghams Gap, Queensland *(Phot by Gordon Grigg).*

SIZE: Males 24–30 mm; females 24–25 mm.

HABITAT: *Kyarranus kundagungan* is a mountain rainforest species with similar habits to *K. loveridgei*. It is usually found among saturated or moist leaf litter and vegetation near small creeks. The distribution is confined to the mountain areas between the Mistake Mountains and Teviot Falls, Queensland.

ADVERTISEMENT CALL: A slow, guttural, low-pitched "ork", very similar to *K. loveridgei*. Calling occurs between August and February.

BREEDING AND LARVAE: Very large yolky eggs are deposited in water-filled burrows. Larvae grow to about 20 mm and are lightly pigmented with narrow fins and a rounded tail tip. Labial teeth are absent.

SIMILAR SPECIES: The dorsal and ventral coloration and smooth dorsum should be sufficient to distinguish adults of this species from other members of the genus.

Kyarranus loveridgei

(Parker)

Loveridge's Frog

The dorsum, which may have a number of small warts, is uniformly dark brown. The ventral surface is pale, usually cream or light orange-brown and the throat is mottled. A dorso-lateral ridge runs from the eye to near the groin. The only conspicuous markings are a dark stripe on the side of the head from the snout to the shoulder and faint cross-barring which may be present on the hind limbs. The fingers and toes are free from webbing and vomerine teeth lie in two oblique rows to the rear of the choanae.

SIZE: Males 22–27 mm; females 29–30 mm.

HABITAT: *Kyarranus loveridgei* occurs in the rainforests of the ranges near the Queensland border and south to the Gibraltar Range. Frogs are found in soft, moist earth at elevations of about 1 000 m and above.

Kyarranus loveridgei, Mt Warning, New South Wales *(Photo by Gordon Grigg)*.

Kyarranus sphagnicolus, Point Lookout, New South Wales *(Photo by Gordon Grigg)*.

ADVERTISEMENT CALL: Males call from underground, emitting a slow, guttural, low-pitched "ork". The recording is of one calling from the opening of a flask-shaped soil burrow 8 cm in depth, under leaf litter beside a small creek at Mount Warning, New South Wales. (Tape track 30.)

BREEDING AND LARVAE: Higher temperatures in November and December probably stimulate breeding in this species. 20–30 eggs are deposited in a frothy mass in a nest dug 8–10 cm into moist earth. As in all terrestrial breeders, the eggs are large and yolky. The tadpoles lack tooth rows.

SIMILAR SPECIES: Confusion of *K. loveridgei* with any other species is unlikely because of its limited range and specialized habits. *K. sphagnicolus* is larger, has a more distinctive dorsal pattern and is generally distributed south of *K. loveridgei*. *K. kundagungan* has distinctive dorsal and ventral coloration.

Kyarranus sphagnicolus

Moore

Sphagnum Frog

This species has the largest range of any species of the genus and is specialized to life in moist rainforests at moderate to high elevations. Its distribution extends as far south as Gloucester Tops and north to Mt Hyland. *Kyarranus sphagnicolus* is a small stout frog. The dorsal background colour is greenish-grey in males and reddish-brown in females. In both sexes there are isolated but prominent warts which tend to form lines along the back. Dark bands joining anteriorly form an inverted V on the posterior of the dorsum. A dark stripe is present on the side of the head, running from the snout to the shoulder. The undersurface is smooth and may be white or pale orange-brown with the throat heavily mottled in dark grey. The thighs are black posteriorly. Fingers and toes lack webbing. In breeding condition, females have spatulate fingers whereas males have small nuptial pads. The vomerine teeth form two oblique series to the rear of the choanae.

SIZE: Males 30–34 mm; females 30–34 mm.

HABITAT: This species is known from scattered areas in the northern and central northern mountains of New South Wales. It is found only in areas of high rainfall (1 500 mm annually) where suitable habitats are to be found. At Point Lookout, where it was first discovered, the Sphagnum Frog lives in the extensive beds of moist sphagnum moss.

ADVERTISEMENT CALL: A resonant hawser-like "creeerk", quite variable. (Tape track 31.)

BREEDING AND LARVAE: 40–55 large eggs are deposited in a frothy mass in moist spots such as in rock crevices, under logs, or in burrows made into sphagnum moss. In aquatic situations, the unpigmented larvae may be found in the silt of small pools. They lack tooth rows. Development is completed in one month.

SIMILAR SPECIES: It is unlikely that this species could be confused with any other frog in the same locality. See also *K. loveridgei* and *K. kundagungan*.

Genus *Taudactylus* Straughan and Lee

The genus *Taudactylus* includes six species, each of which is confined to rainforest creeks in eastern Queensland. *Taudactylus* is distinguished from other myobatrachid genera by the T-shape of the terminal phalangeal bone of each finger and toe. This feature is not unique, for T-shaped bones characterize some *Platymantis* species of the south-west Pacific (Ranidae), whilst T- and Y-shaped bones are found in several microhylid genera, including *Oreophryne, Sphenophryne* and *Cophixalus*.

Most *Taudactylus* species have undergone a reduction in range and are of conservation concern, being classified as "endangered" or "vulnerable". One (*T. diurnus*) may be extinct.

Taudactylus species are small to moderate sized frogs, rarely exceeding 35 mm in length. The skin is smooth and the fingers and toes are unwebbed. The tympanum is large and visible or hidden. They lack vocal sacs, but most species are known to vocalize.

Key to species of *Taudactylus* (after Czechura 1986)

1a. A pale, dorso-lateral skin fold from eye to groin .... *T. acutirostris* (p. 251)
b. No dorso-lateral skin fold from eye to groin 2
2a. Discs on fingers and toes distinctly wider than penultimate phalanx 3
b. Discs on fingers and toes only slightly wider than penultimate phalanx 4
3a. Yellowish-green above with distinct dark markings; ventral surface of thighs immaculate yellow .. *T. eungellensis* (p. 255)
b. Dark olive-brown to blackish with identical markings; ventral surface of thighs greyish-cream spotted with dark grey .... *T. diurnus* (p. 253)
4a. Continuous dark lateral band from eye to groin .... *T. rheophilus* (p. 258)
b. Dark lateral band disrupted above insertion of arm 5
5a. Broad interorbital bar; ventral surface with intense brown speckling .... *T. pleione* (p. 258)
b. Triangular interorbital patch; ventral surface with pale speckling .... *T. liemi* (p. 257)

Taudactylus acutirostris

(Andersson)

Sharp-nosed Torrent Frog, Tinker Frog, Sharp-snouted Day Frog

The most distinctive member of the genus. The dorsum is pale to dark brown with a greyish or yellowish tinge. The sides are dark grey or black. These two areas are separated by a pale dorso-lateral fold of skin. The ventral surface is cream to pale reddish-brown. Dorsal surfaces of the limbs are cross-barred. The snout is acutely pointed and projects well beyond the lower jaw. The side of the head is dark and there is a thin pale stripe from the tip of the snout to the eye. Fingers and toes are free from web, and the discs are distinct.

SIZE: Males 22–27 mm; females 29–30 mm.

HABITAT: Found at the margins of montane creeks in north-east Queensland.

ADVERTISEMENT CALL: Two different call types are known. In one, the calls have a duration of 4–6 seconds and are highly pulsed, calls being repeated about seven times each minute and resemble "eek...eek...eek...eek...". In the other, the call is a metallic "tink...tink...tink..." somewhat like a glass being tapped with a spoon. Males call from exposed positions on rocks or from adjacent leaf litter.

BREEDING AND LARVAE: Similar to *T. diurnus* except that tadpoles have a tooth formula of $\frac{1\;^{1}\;1}{1\;_{2}\;1}$ and the marginal papillae are interrupted anteriorly at the extremities of the outer tooth row.

SIMILAR SPECIES: *Taudactylus acutirostris* is easily distinguished from other Australian frogs by the acutely pointed snout and contrasting dorsum and lateral surfaces separated by the pale dorso-lateral skin fold.

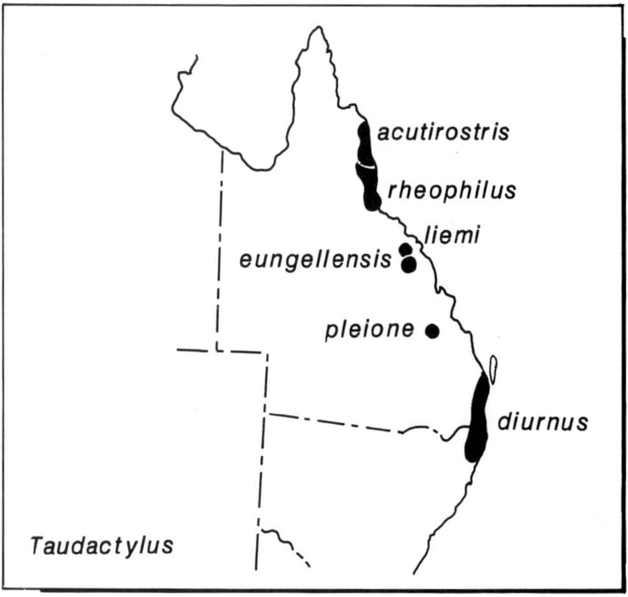

Taudactylus acutirostris, Tinaroo Creek Road, north Queensland *(Photo by Gordon Grigg)*.

Taudactylus diurnus

Straughan and Lee

Mt Glorious Torrent Frog, Southern Day Frog

The dorsal background colour is greyish brown with darker spotting. The most consistent marking is an H between the shoulders, with a dark bar between the eyes. The ventral surface is cream or pale yellow with large light grey spots. The limbs have dark cross-bars on the dorsal surface and are heavily mottled on the ventral surface. The hands and feet are white with black spots. Vomerine teeth are absent, and the tympanum is hidden. The discs on the fingers and toes are quite distinct and have a longitudinal groove. Lateral fringes on the toes are also conspicuous.

SIZE: Males 22–27 mm; females 22–31 mm.

HABITAT: *Taudactylus diurnus* is (or was?) known from three rainforest areas in southeastern Queensland: Mount Glorious, the Conondale Ranges and Kondalilla (3 km SW of Montville). Found in and around pools and flowing streams by day as well as night, they are active little frogs spending much of their time in the water.

Taudactylus diurnus, Mt Glorious, Queensland *(Photo by Gordon Grigg)*.

Taudactylus diurnus, ventral view, Kondalilla, Queensland *(Photo by Gordon Grigg)*.

ADVERTISEMENT CALL: A soft chuckling sound repeated in short bursts. The soft sound is attributable to the fact that males lack vocal sacs.

BREEDING AND LARVAE: Eggs are large and are usually found adhering to the underside of rocks or to submerged branches. The tadpoles are dark brown with a globose body and narrow fins. The unique characteristic is the umbrella-shaped oral disc, completely surrounded by papillae and without any tooth rows. The breeding season extends from November to May.

SIMILAR SPECIES: The most similar species is *T. eungellensis* but, because of the wide geographic separation of that species from *T. diurnus*, the locality data will prevent confusion. Amongst other south-east Queensland frogs there is a superficial resemblance to *Crinia* species, but these are not active in the day (unless disturbed from their resting place) and they lack the expanded discs on the tips of the digits.

Taudactylus eungellensis

Liem and Hosmer

Eungella Torrent Frog, Eungella Day Frog

Taudactylus eungellensis is the only Australian Myobatrachid known to advertise its presence by movements of its body and limbs. Whereas the four known Australian species appear to be associated with aggression by one male to another, we suspect that the display of *T. eungellensis* is associated with courtship. The head and body are slender, the limbs long and lean. The back is dull brown or black dappled with numerous tan or orange patches. There is an X-shaped mark on the back. The undersurface is deep yellow.

SIZE: Males 25–28 mm; females 28–36 mm.

HABITAT: Confined to closed rainforest in the Clarke Range, Finch Hatton Gorge west of Mackay in north-east Queensland. The frogs inhabit the floor of rock-strewn creek beds.

Taudactylus eungellensis, Eungella National Park, north Queensland *(Photo b Jean-Marc Hero).*

Taudactylus liemi, Eungella, north Queensland *(Photo by Steve Richards).*

ADVERTISEMENT CALL: A very soft tinkering sound, barely audible above the sound of cascading water.

BREEDING AND LARVAE: Egg masses of 30–50 eggs are laid under rocks in water in November or December. Eggs are 2.2–2.6 mm diameter. The mouth disc is shaped roughly like an umbrella with the outer margin loose and projecting as a flap. The horny beak is small and there are no labial tooth rows. The tadpoles are found on the floor of rock pools or fast-flowing creeks.

SIMILAR SPECIES: Within its limited area of distribution there is only one other member of the genus, *T. liemi*, which has very small digital discs (not distinctly expanded) and a greyish-brown dorsal colouration lacking the X-shaped marking of *T. eungellensis*.

Taudactylus liemi

Ingram

Liem's Frog, Eungella Tinker Frog

This is a secretive species characterized by having very small discs on the tips of the slender fingers and toes. The undersurface of the body bears extensive brown mottling on a cream background. The dorsal surface is a greyish-brown, with a darker brown, triangular patch on the head and a lyre-shaped marking on the back.

SIZE: Males 21–29 mm; females 28–29 mm.

HABITAT: Confined to the rocky margins of fast flowing creeks in montane rainforest in the Eungella area west of Mackay, from Mt William in the north to Crediton in the south.

ADVERTISEMENT CALL: A short series of "tinks" repeated at a rate of about 25 times per minute. Males call from hidden positions.

BREEDING AND LARVAE: The egg clumps and tadpoles of this species remain unknown. Dissections of preserved, gravid females reveal an egg complement of 34–51, each egg measuring 1.7–2.5 mm.

SIMILAR SPECIES: Of the *Taudactylus* species with small discs on the fingers and toes, the dorsal markings (triangular head patch and lyre-shaped marking on the back) provide a simple means of distinguishing *T. liemi* from all others.

Taudactylus pleione

Czechura

Pleione's Torrent Frog, Kroombit Tinker Frog

The dorsum is grey or bluish-grey with darker markings including a bar between the eyes separated from an X-shape or hourglass shape in the middle of the back. There are dark patches in the groin and the undersurface is speckled with dark grey or brown.

SIZE: Adult males 25–27 mm; females 26–28 mm.

HABITAT: This species is confined to the headwaters of Kroombit Creek on Kroombit Tops in south-east Queensland, where it lives amongst leaf litter or rocks adjacent to permanent pools or running water.

ADVERTISEMENT CALL: Described as a series of rapidly repeated metallic "tinks" emitted at regular intervals.

BREEDING AND LARVAE: Unknown.

SIMILAR SPECIES: The highly localized distribution of *T. pleione* prevents confusion with any other species, being geographically isolated from them.

Taudactylus rheophilus

Liem and Hosmer

Tinkling Frog, Northern Timber Frog

The dorsum is reddish-brown to dark brown with darker markings, the most consistent being a dark bar between the eyes. Lateral surfaces dark grey or black. Like *T. acutirostris,* these contrasting areas are separated by a pale dorso-lateral line although it is not as conspicuous and not in the form of a ridge. This stripe does not continue forward to the snout. The ventral surface is yellowish and the limbs have dark crossbars. The snout is rounded and does not project

audactylus pleione, Kroombit Tops, Queensland *(Photo by Owen Kelly).*

audactylus rheophilus, Mt Lewis, north Queensland *(Photo by Owen Kelly).*

beyond the lower jaw. As in all members of the genus, vomerine teeth are absent and the tympanum is hidden. The discs on fingers and toes are conspicuous and webbing is absent.

SIZE: Males 24–27 mm; females 24–31 mm.

HABITAT: *Taudactylus rheophilus* is known only from an altitude of 1 200 m on the south-west slopes of Mount Lewis, Thornton Peak and the Lamb Range. The streams in which it has been found are extremely fast flowing.

ADVERTISEMENT CALL: A soft metallic tapping note repeated 4–5 times in quick succession. The only observation was of a male calling between two rocks with the body submerged and only the head above the water.

BREEDING AND LARVAE: The form of the egg mass and tadpole has not been observed, but eggs dissected from preserved gravid females ranged in number from 35–50 and were 1.8–2.4 mm diameter.

SIMILAR SPECIES: The extremely sharp snout and distinct lateral ridge distinguish this species from all other *Taudactylus*. Lack of webbing between the toes will distinguish it from juvenile *Rana daemeli* which may look superficially similar.

Genus *Metacrinia* Parker

The single species of *Metacrinia* is restricted to a narrow coastal zone in the extreme south-west of Western Australia. The recognition of the genus as distinct from *Pseudophryne* (to which it is very closely related), hinges upon its possession of a suite of biological and structural features, including vivid markings on the ventral surface, and a pattern of larval development entirely out of water.

Metacrinia nichollsi

(Harrison)

Nicholls' Toadlet

The body is squat and has short limbs and short and unwebbed fingers. The tympanum is visible. The skin is extremely warty above and coarsely granular beneath. The dorsal coloration is a very dark brown without markings. Beneath, the overall coloration ranges from deep grey through slate to deep purple, highlighted by small yellow or vivid orange patches adjacent to the limbs and on the underside of the thighs and calves.

SIZE: Males 19–21 mm; females 22–24 mm.

HABITAT: Found principally in leaf litter and beneath rotting logs in the Karri forests.

ADVERTISEMENT CALL: A short "ark" resembling the call of *Pseudophryne* species.

BREEDING AND LARVAE: In late summer 25–30 eggs are laid in leaf litter or beneath other moist ground cover. Details of the tadpoles have not been reported.

SIMILAR SPECIES: Likely to be confused only with *Pseudophryne guentheri* which is the only toadlet found in the same area as *M. nichollsi*. *Pseudophryne guentheri* is a larger species (26–33 mm) and lacks the ventral orange or yellow markings of *M. nichollsi*.

Metacrinia nichollsi, Walpole, Western Australia *(Photo by Gordon Grigg*
(Specimen courtesy Grahame Webb).

Metacrinia nichollsi, ventral view, Walpole, Western Australia *(Photo by Gordon*
Grigg) (Specimen courtesy Grahame Webb).

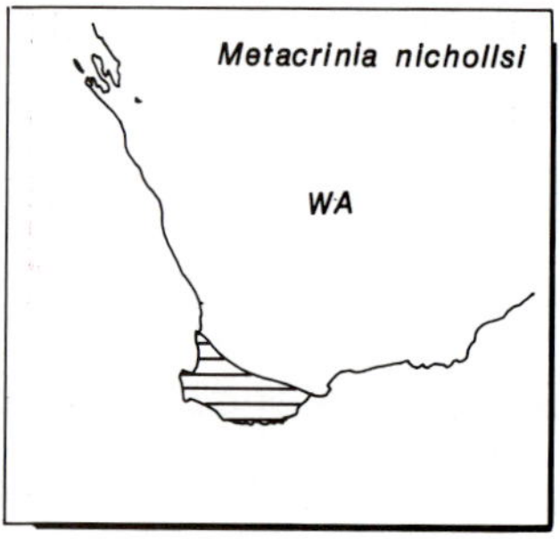

Genus *Pseudophryne* Fitzinger

This genus is composed of seven southeastern and three Western Australian species. All are small terrestrial frogs, or "toadlets" as they have been popularly called, that are similar in body shape and size (maximum length is 30 mm). The mating calls of all species are remarkably similar and are of little help in identification. However, all species are distinguishable by their colour pattern, which may be surprisingly vivid, considering their secretive habits. There are differences of opinion regarding the taxonomic status of *Pseudophryne bibronii, P. major* and *P. coriacea* in south-east Queensland and northern New South Wales and the status of *P. bibronii* and *P. dendyi* in southern New South Wales.

The habits of all species seem to be very similar, and they are most commonly found under logs or stones in damp, marshy areas. *Pseudophryne corroboree* is typically found in sphagnum bogs or wet heath and *P. australis* is often found under rocks near seepages in sandstone country.

The breeding biology of this group is of considerable interest because pre-mating isolation seems to break down in some areas of geographic overlap and hybrids are commonly found.* The loss of the outer and part of the inner ear is interesting in this connection. How this group of frogs hear airborne sounds remains unsolved. Eggs are usually deposited out of water in moist earth. The eggs are large and fewer in number than those of fully aquatic species. After developing to the hatching stage, the larvae may enter a period of dormancy until sufficient rain falls to either flood the immediate area or wash the eggs into a creek. Considerable variation in rates of development has been observed. *Pseudophryne douglasi* of Western Australia is the only aquatic breeding species.

---

*Hybrids will show characteristics of both species to varying degrees. Obviously, it is more difficult to key hybrid specimens, especially if the reader has not seen typical examples of both species. The commonest hybrids found are *P. bibronii* × *P. semimarmorata*, *P. bibronii* × *P. coriacea*, and *P. bibronii* × *P. dendyi*.

Key to *Pseudophryne* species

1a. Ventral surface of limbs and throat pale tan to orange .... *P. semimarmorata* (p. 279)
b. Ventral surface of limbs black and white 2

2a. Two large transverse metatarsal tubercles. Inner toe with only one phalanx .... *P. guentheri* (p. 274)
b. Outer metatarsal tubercle not transversely compressed, inner larger than outer. Two phalanges in first toe 3

3a. An oval gland at the distal end of the posterior of the thigh 5
b. No glands on thigh 4

4a. Dorsum with small conical warts .... *P. douglasi* (p. 274)
b. Dorsum with flattened warts .... *P. occidentalis* (p. 277)

5a. Snout pointed, nostrils directed vertically upwards. Dorsum with longitudinal rows of dark warts. Maximum body length 36 mm. .... *P. major* (p. 277)
b. Snout rounded. Nostrils dorso-lateral. Dorsum without longitudinal rows of dark warts. Body length to 34 mm 6

6a. Dorsum with bright yellow or pale green and black stripes .... *P. corroboree* (p. 270)
b. Dorsum without yellow or green and black stripes 7

7a. Dorsum reddish-brown to red. Sides very dark .... *P. coriacea* (p. 270)
b. Dorsum other than above 8

8a. A reddish T-shaped mark on head. A reddish stripe on urostyle. A white spot on upper arm .. *P. australis* (p. 264)
b. Not as above 9

9a. Dorsum grey to dark brown. A pale yellow stripe on urostyle. A pale yellow spot on upper arm and beneath cloaca .... *P. bibronii* (p. 267)
b. Dorsum very dark brown to jet black. Bright yellow marks on urostyle, upper arm, and beneath cloaca .... *P. dendyi* (p. 272)

Pseudophryne australis

(Gray)

Red-crowned Toadlet

The common name aptly describes this little toadlet, which is found only in the Hawkesbury Sandstone country around Sydney.

The background colour of the dorsum varies from chestnut to dark brown with red, black and white spots scattered over the surface. The red "crown" consists of a red or orange T-shaped patch on the head,

Pseudophryne australis, Menai, New South Wales *(Photo by Marion Anstis)*.

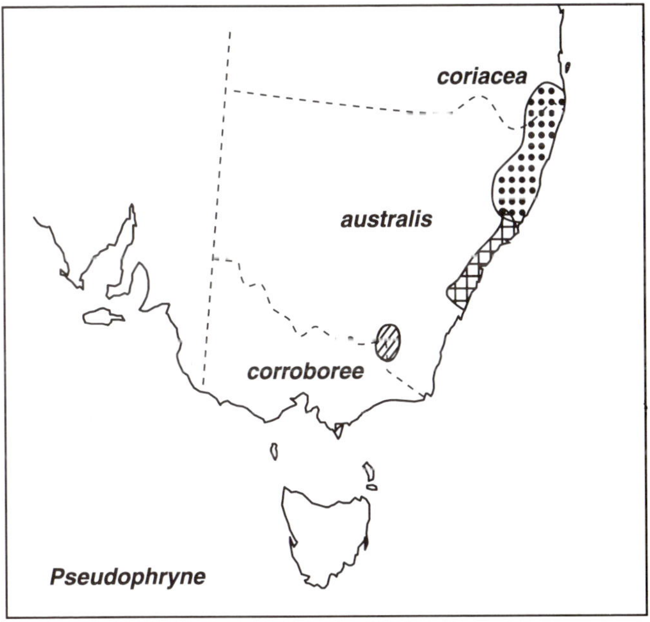

with the horizontal bar joining the eyes and the vertical portion extending to the tip of the snout. Another prominent red stripe is usually present on the urostyle. The lateral surfaces may be a deep dapple-grey while the underparts are marked in typical *Pseudophryne* pattern. There is a conspicuous white spot on the upper arm.

SIZE: Males 22–28 mm; females 25–29 mm.

HABITAT: *Pseudophryne australis* is found in a variety of damp situations on the Hawkesbury Sandstone areas from the Royal National Park in the south to between Gosford and Newcastle in the north. The species is often gregarious and colonies may number between 20 and 30 individuals. One of the unusual places in which specimens have been found was in deep crevices in a sandstone rock face. Water seepages had eroded the inner edge of the crevices to form a labyrinth of caverns and tunnels where a large number of toadlets were breeding, their creaking calls reverberating through the miniature cave system.

ADVERTISEMENT CALL: As for *P. bibronii*.

BREEDING AND LARVAE: About 20 large eggs are deposited in a crude nest. Males have been observed in attendance. Development is much more rapid than in *P. bibronii*, the larvae taking two to three months to complete metamorphosis.

SIMILAR SPECIES: In the 1830s when this species was first described, many taxonomists apparently confused it with *P. bibronii*. This may be due to an error in the type locality, given as Swan River, Australia. Even in the 1870s Fletcher, a leading herpetologist, considered that *P. australis* was a subspecies of *P. bibronii*. It is difficult to see how confusion could occur as consistent differences are readily recognized. *Pseudophryne bibronii* is not commonly found in the Hawkesbury Sandstone country.

Pseudophryne bibronii

Günther

Bibron's Toadlet, Brown Toadlet

Pseudophryne bibronii was the most common and widespread member of the genus but populations appear to have declined in some areas. The dorsum is pale to dark brown with irregular flattened warts. Black, red, or white dots may be scattered over the back and there is often a paler stripe on the urostyle. The ventral surface is boldly marbled in black and white. A pale yellow patch surrounds the cloaca and may even cover the posterior of the thigh. A similar coloured patch occurs on the upper arm. The arms and legs are very short with unwebbed toes and fingers. Vomerine and maxillary teeth are lacking as in all members of the genus.

SIZE: Males 22–30 mm; females 25–32 mm.

HABITAT: The Brown Toadlet is found in situations where there is sufficient damp soil and cover. Where there is an absence of fallen logs or stones, they have been found in long tunnels about 3 cm under the soil.

ADVERTISEMENT CALL: Probably the nearest description of the call is a short, grating, upwardly inflected "ark".

BREEDING AND LARVAE: Calling commences in February and continues through the winter to August. Temperatures recorded during calling have been as low as 4°C. Large eggs are deposited in damp leaf mould, under logs or stones. Hatching takes place after sufficient rain floods the area. Tadpoles reach a maximum of 31 mm and are uniformly dark brown with a mottled tail fin. Metamorphosis occurs from three to seven months after hatching. The tooth formula is $\frac{1\ ^{1}\ 1}{1\ _{2}\ 1}$.

SIMILAR SPECIES: Most members of this genus are quite distinct and confusion should not occur. However, in southern New South Wales, *P. bibronii* may be confused with *P. dendyi* which has similar markings, but the yellow is much brighter and the dorsum much darker in *P. dendyi*. *Pseudophryne semimarmorata* can be distinguished by the tan coloration on the ventral surfaces of the limbs and the extensive yellow, ventral pattern. *Pseudophryne major* is distinguished by its large size and pointed snout.

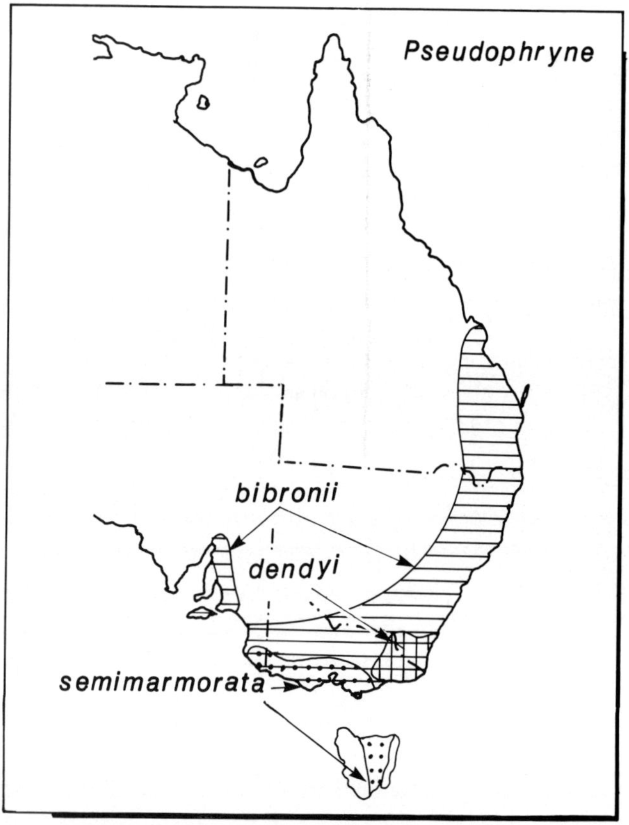

seudophryne bibronii, Flinders Ranges, South Australia *(Photo by Mike Mahony).*

seudophryne coriacea, Tenterfield, New South Wales *(Photo by Gordon Grigg).*

Pseudophryne coriacea

Keferstein

Keferstein's Toadlet, Red-backed Toadlet

Less is known about this species than any other in the genus. The dorsal surface varies from rusty brown to a deep red and separating this bright colour from the marbled ventral surface is a broad lateral black band from the snout to the groin. Specimens from the far northern coast of New South Wales are relatively sombre and may represent hybrids with *P. bibronii*.

SIZE: Males 19–24 mm; females 23–27 mm.

HABITAT: Fallen bark and leaf litter in moist areas seem to be the preferred habitat of Keferstein's Toadlet. The dorsal colour is magnificently camouflaged to represent similarly coloured leaves in the toadlets habitat.

ADVERTISEMENT CALL: Same as *P. bibronii*.

BREEDING AND LARVAE: About 80 eggs are deposited in moist earth and development takes about two months. The larvae are indistinguishable from those of *P. bibronii*.

SIMILAR SPECIES: Unlikely to be confused with any other species except in the hybrid zone of northern New South Wales.

Pseudophryne corroboree

Moore

Corroboree Frog

Of all the frogs in Australia, the Corroboree Frog must surely be the most unusual and attractive. Therefore it is surprising that it was not named until 1953 by John Moore, although the first specimen was presented to the Australian Museum in 1947. There are two colour forms. The "southern" form has the characteristic bright yellow and black stripes seen in most photographs of the species. The "northern" form has reduced narrow lime-green or yellow stipes which are frequently broken up with the black stripes. The ventral surface is similar to other *Pseudophryne* species except that in some specimens the pale areas may be yellow, yellow and pale blue, or white.

ꞏeudophryne corroboree, Smiggins Holes, New South Wales *(Photo by Gordon Grigg) (Specimen courtesy Dane Wimbush).*

SIZE: Males 26–28 mm; females 26–30 mm.

HABITAT: Corroboree Frogs are found in the Southern Alps above 1 000 m where there are extensive sphagnum bogs. At altitudes below the tree line, fallen logs and debris may be used for shelter.

ADVERTISEMENT CALL. Similar to *P. bibronii*, but more drawn out.

BREEDING AND LARVAE: Due to the climate there is only a short breeding season, from January to March. Nests are made in deep burrows in sphagnum moss, grass or sedges where 10 to 38 large eggs are deposited. Males have been observed staying with the eggs. Tadpoles are very similar to other *Pseudophryne* species, although the colour pattern is visible as metamorphosis approaches.

SIMILAR SPECIES: This species is so remarkably distinct that confusion with other species is impossible.

Pseudophryne dendyi

Lucas

Southern Toadlet

The validity of this species has been in doubt for many years. It is closely related to *P. bibronii* and a number of herpetologists are of the opinion that *P. dendyi* is a sub-species or geographical race of the Brown Toadlet. Specimens from above 1 300 m in the Snowy Mountains could be considered "typical" *dendyi*. They are very dark brown-black on the dorsum, with a bright yellow stripe on the urostyle and with similarly coloured patches on the cloaca and posterior of the thigh (often extending down to the toes). A few specimens have irregular yellow patches on the head. The ventral surface is marbled in black and white.

SIZE: Males 24–28 mm; females 25–29 mm.

HABITAT: Found in similar situations to *P. bibronii*.

ADVERTISEMENT CALL: Same as *P. bibronii*. (Tape track 28).

BREEDING AND LARVAE: As for *P. bibronii*.

SIMILAR SPECIES: Confusion may occur between *P. dendyi* and *P. bibronii*. Usually the locality is sufficient, although this is a poor criterion to depend on for species recognition. It is reasonable to say that any specimen which is almost black on the dorsum and has bright and extensive yellow patches on the thigh is *P. dendyi*, and individuals with less and paler yellow marking are *P. bibronii*.

eudophryne dendyi, Thurra River, Victoria *(Photo by Gordon Grigg).*

eudophryne dendyi, ventral view, Thurra River, Victoria *(Photo by Gordon Grigg).*

Pseudophryne douglasi

Main

Douglas's Toadlet

The dorsum is chocolate brown with an orange triangle on the snout, a similar mark pointing to the rear on the urostyle and a few small orange blotches on the dorsum. The skin has numerous small conical warts. The belly is marbled in black and white. Fingers and toes are short and free from web. There is a large transverse inner metatarsal tubercle and a small circular outer tubercle.

SIZE: Males 23–28 mm; females 27–32 mm.

HABITAT: *Pseudophryne douglasi* has a specialized habitat preference. It is found in shallow streams and creeks, unlike other *Pseudophryne* which shun the water. They have been found in burrows under stones at the water's edge and leap readily into the water, swimming with both legs together.

ADVERTISEMENT CALL: Similar to *P. bibronii*.

BREEDING AND LARVAE: Large eggs are deposited in water under stones, a behaviour which is unlike all other *Pseudophryne*. The larvae are very similar to other species of this genus.

SIMILAR SPECIES: *Pseudophryne douglasi* is geographically separated from all other *Pseudophryne*. The only other small genus found in the same localities is *Uperoleia*, from which it can be distinguished by the absence of red, orange or yellow markings in the groin and on the posterior of the thigh.

Pseudophryne guentheri

Boulenger

Günther's Toadlet

A common and distinctive species from the south-west of Western Australia, *P. guentheri* grows to a maximum body length of about 30 mm. The dorsum is pale grey or brown with darker spots and blotches. The skin has numerous small warts and there are two converging folds between the shoulders giving the appearance of

seudophryne douglasi, Python Pool, Western Australia *(Photo by Steve Donnellan).*

seudophryne guentheri, Walpole, Western Australia *(Photo by Gordon Grigg) (Specimen courtesy Grahame Webb).*

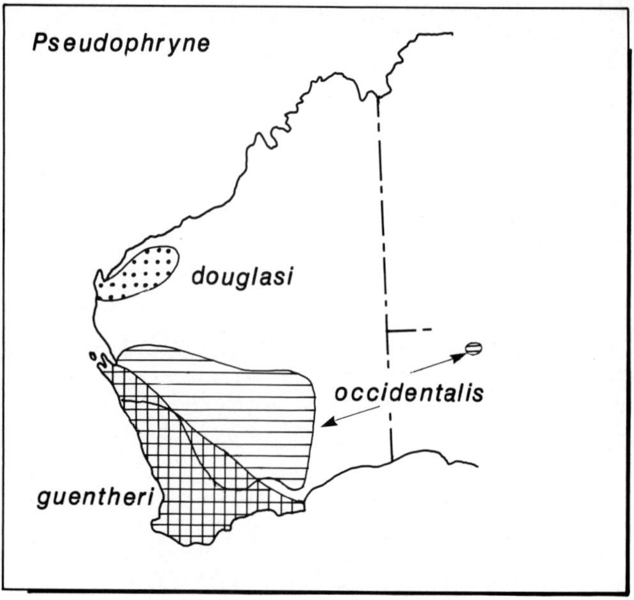

an X or > <. There is often a pale patch on the head and on the parotoid region of each shoulder, with a broad pale stripe on the urostyle. The belly is typical of the genus. Toes are short and fleshy, the first having a single phalanx. A very distinctive feature is the presence of two large transversely compressed tubercles on each foot. The parotoid glands are fairly distinct and there is another gland visible in the groin. Post-femoral glands are absent.

SIZE: Males 26–30 mm; females 29–33 mm.

HABITAT: Similar to *P. bibronii*.

ADVERTISEMENT CALL: Similar to *P. bibronii*.

BREEDING AND LARVAE: An autumn breeder; otherwise the life cycle is similar to *P. bibronii*.

SIMILAR SPECIES: *P. guentheri* is distinguished from other *Pseudophryne* in the south-west of Western Australia by the single phalanx in the first toe (all other species have two phalanges), and the two transverse metatarsal tubercles (in other species, the outer tubercle is small and circular).

Pseudophryne major

Parker

Large Toadlet

Little is known about this large species from eastern Queensland. The dorsum is grey with a broad richly brown area mid-dorsally with darker blotches and several rows of flattened warts arranged longitudinally and a pale stripe along the urostyle. There is a large pale spot at the base of the upper arm. The belly is typical of the genus. The snout is pointed, with the nostrils situated on the dorsal surface. There is a distinct inner and a small outer metatarsal tubercle.

SIZE: Males 24–29 mm; females 27–31 mm.

HABITAT: *P. major* occurs widely in eastern Queensland to north-east New South Wales. It is found in similar boggy soakages as other species of the genus.

ADVERTISEMENT CALL: Similar to *P. bibronii.*

BREEDING AND LARVAE: Similar to *P. bibronii.*

SIMILAR SPECIES: See *P. bibronii.*

Pseudophryne occidentalis

Parker

Orange-crowned Toadlet

This species is adapted to an arid environment, and is found in the very dry areas of northwestern South Australia and southern Western Australia. Dorsal colour is grey to chocolate brown with a triangular orange patch on the head, an orange stripe on the urostyle, and a pale orange spot on the upper arm. The skin may be smooth or have numerous flattened warts. The ventral surface is marbled with black and white. The parotoid glands may be distinct but there are no glands in the groin or on the posterior of the thigh.

SIZE: Males 22–24 mm; females 21–26 mm.

HABITAT: *Pseudophryne occidentalis* is found in shallow tunnels or under debris or rocks near temporary or permanent pools.

Pseudophryne major, Gayndah, Queensland *(Photo by Steve Donnellan).*

Pseudophryne occidentalis Paynes Find, Western Australia *(Photo by Stev Donnellan).*

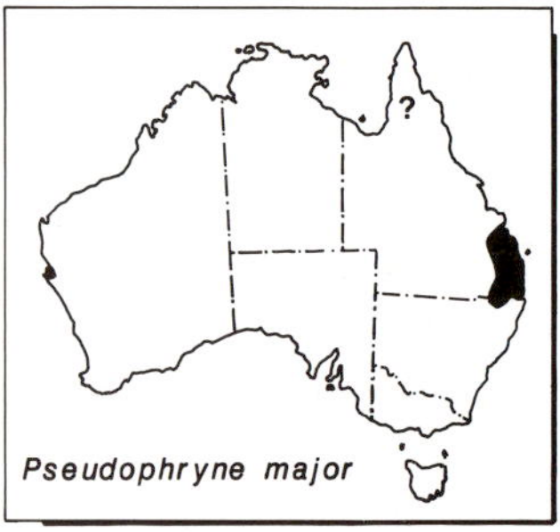

ADVERTISEMENT CALL: Similar to *P. bibronii*.

BREEDING AND LARVAE: Similar to other *Pseudophryne*, *P. occidentalis* breeds during summer and autumn (December to May).

SIMILAR SPECIES: This species can be distinguished from *P. douglasi* by the flattened rather than conical warts and from *P. guentheri* by the lack of folds above the shoulders.

Pseudophryne semimarmorata

Lucas

Southern Toadlet

The dorsum varies from chocolate-brown to dark olive-green with numerous distinct warts. The ventral surface is similar to other *Pseudophryne* species except for the throat and undersides of the limbs which are pale tan to orange. Fingers and toes are short, without webbing.

SIZE: Males 25–28 mm; females 25–33 mm.

HABITAT: Similar to *P. bibronii*. This is the only *Pseudophryne* found in Tasmania and on the Bass Strait islands.

ADVERTISEMENT CALL: Similar to *P. bibronii*.

BREEDING AND LARVAE: *Pseudophryne semimarmorata* breeds during autumn. Otherwise similar to *P. bibronii*.

SIMILAR SPECIES: See *P. bibronii*.

Pseudophryne semimarmorata, Bairnsdale, Victoria *(Photo by Steve Donnellan)*.

Pseudophryne semimarmorata, frogs with egg mass in nest under a rock, Moondarra, Victoria *(Photo by Murray Littlejohn)*.

Genus *Paracrinia* Heyer and Liem

Paracrinia was established in 1976 to accommodate the species *haswelli*, which had previously been included in *Crinia*.

Paracrinia haswelli

(Fletcher)

Haswell's Froglet

The head and body are slender, the legs slightly longer than in other froglets, and the toes are long and narrowly fringed. The skin of the back is smooth but for a few scattered tubercles. The coloration is dull brown with a few darker spots, and a broad, black or intense dark brown stripe from the eye to above the insertion of the forearm. Most specimens have a pale mid-vertebral stripe. On the hidden surfaces of the hind limbs are spots of brick red.

SIZE: Males 31–33 mm; females 32–35 mm.

HABITAT: Found at the edge of swamps, marshes and pools in a narrow coastal strip from near Sydney, New South Wales to the extreme east of Victoria.

ADVERTISEMENT CALL: A harsh "wronk" (Tape track 9).

BREEDING AND LARVAE: Details of breeding and development have not been published. Eggs are laid in permanent water. The tadpole tooth formula is $\frac{1\ ^{1}\ 1}{2}$.

SIMILAR SPECIES: Other small frogs that might be mistaken for *P. haswelli* are *Crinia signifera* and *Geocrinia victoriana*. *Crinia signifera* lacks the lateral dark head stripe and has no brick red spots on the limbs. *Geocrinia victoriana* has pink spots.

Paracrinia haswelli, Smiths Lake, New South Wales *(Photo by Gordon Grigg)*.

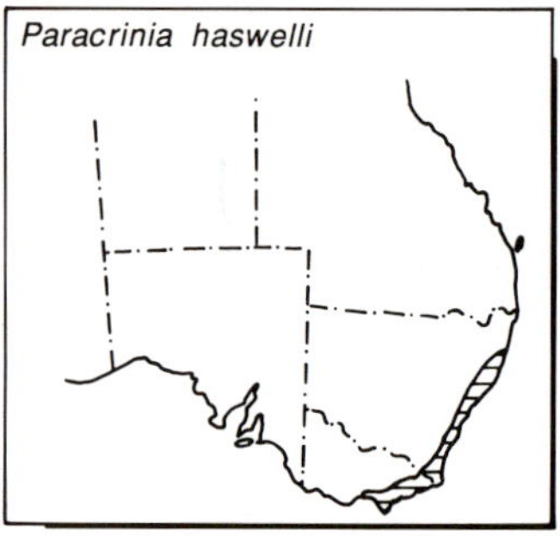

Genus *Crinia* Tschudi

Among Australia's anuran fauna is a large number of species which have in common small body size and many forms of dorsal patterns. Most of these species were, at some time in their taxonomic history, placed in the genus *Crinia* which was created for the brightly coloured species *C. georgiana* of southwestern Western Australia. Many of these species were thought to be races of the species *C. signifera*.

In the 1950s, 60s and 70s a great deal of taxonomic work was undertaken, particularly by A. R. Main and M. J. Littlejohn who described a number of new species using mating call analysis and hybridization experiments. Other studies began splitting the genus into several new genera. *C. darlingtoni* was removed (because it displayed unique parental care) and placed in the new genus *Assa*. *Crinia acutirostris* was transferred to another newly described genus *Taudactylus*. The major rearrangement came in 1973 when *C. signifera* and all its siblings, or similar species, were placed in the genus *Ranidella*. Some were allocated the name *Geocrinia* while *C. georgiana* and *C. haswelli* were left with the original name.

More recently biochemical studies have suggested a closer relationship between *Ranidella* and *Crinia* than their external appearances suggest, with the result that the *Ranidella* species were reinstated to *Crinia*. Here we adopt that broader definition and so recognize two groupings within the genus: the moderate-sized *C. georgiana* with brilliant red patches in the groin and on the thighs, and the smaller and highly variable former *Ranidella* species that lack the red markings.

Crinia (in the broad sense) has long and unwebbed fingers and toes, smooth or granular skins and a tympanum is present or absent. All species lay their eggs singly in small clumps in water. Almost all species are polymorphic — having several variants of colour and skin texture in each population.

Key to *Crinia* species of southwestern Australia

1a. Red patches in groin and on sides of thighs; eyelid red or golden; hands pale .... *C. georgiana* (p. 288)
 b. No red in groin or on thighs; eyelid not red or golden; hands not pale 2

2a. Call resembles a squelch 3
 b. Call does not resemble a squelch 4

3a. Species restricted to Swan R. coastal plain .... *C. insignifera* (p. 289)
 b. Species restricted to forested areas in extreme south of southwestern Western Australia .... *C. subinsignifera* (p. 299)

4a. Call resembles a bleat; throat of male not entirely black .... *C. pseudinsignifera* (p. 293)
 b. Call a rattling sound like a dried pea shaken in a can; throat of male entirely black .... *C. glauerti* (p. 289)

Key to *Crinia* species of northern Australia (Kimberley Division of Western Australia, the entire Northern Territory and Queensland north of Mackay)

1a. Tympanum visible .... *C. bilingua* (p. 284)
b. Tympanum hidden or scarcely detectable 2

2a. Advertisement call long and multi-pulsed .... *C. remota* (p. 293)
b. Advertisement call with two pulses .. *C. deserticola* (p. 286)

Key to *Crinia* species of southeastern Australia (Queensland south of Mackay, New South Wales, Victoria, South Australia, the ACT and Tasmania)

1a. Tympanum visible 2
b. Lacking a tympanum .... *C. riparia* (p. 295)

2a. Patches of pink or red on undersurface of body and limbs .... *C. tasmaniensis* (p. 299)
b. Undersurface lacking pink and red patches 3

3a. White spots on midline of throat .... *C. tinnula* (p. 301)
b. Lacking white spots on midline of throat 4

4a. Advertisement call a long, high-pitched series of clicks .... *C. signifera* (p. 295)
b. Advertisement not a series of clicks 5

5a. Largely confined to arid areas .... *C. deserticola* (p. 286)
b. Does not occur in seasonally arid areas 6

6a. Advertisement call resembles a wet finger being drawn across an inflated balloon .... *C. parinsignifera* (p. 291)
b. Advertisement call comprises a series of short, metallic clicks .... *C. sloanei* (p. 297)

Crinia bilingua

Martin, Tyler and Davies

Bilingual Froglet

A small, short-limbed species with an ornate pattern of markings including a pale bar between the eyes preceding a backwardly-projecting triangular patch. The flanks are commonly paler than the dorsal surface. The skin texture includes smooth, warty and lyrate forms. The toes are narrowly fringed and the tympanum is clearly visible.

SIZE: Males 16–23 mm; females 17–20 mm.

HABITAT: Closely associated with marginal vegetation of temporary swamps and creeks.

Crinia bilingua, Dead Horse Springs, Western Australia *(Photo by Bohdan Stankowich-Janusch)*.

ADVERTISEMENT CALL: As indicated by the specific name, *C. bilingua* has two distinct call phases: a short call of 5–7 pulsed notes of above 450 msec duration, changing to trills of about 700 msec duration.

BREEDING AND LARVAE: Spawn is laid in small clumps attached to aquatic vegetation. Tadpoles become dark brown above and off-white beneath. Metamorphosis can take place in as little as 13–14 days and the tooth formula is $\frac{1\ ^{1}\ 1}{1\ _{2}\ 1}$.

SIMILAR SPECIES: *Crinia bilingua* is related closely to (and has been mistaken for) *C. remota*. Morphologically the species can be distinguished by the condition of the tympanum (clearly visible in *C. bilingua*, but not apparent in *C. remota*), by size and features of the advertisement call that are not distinguishable to the ear.

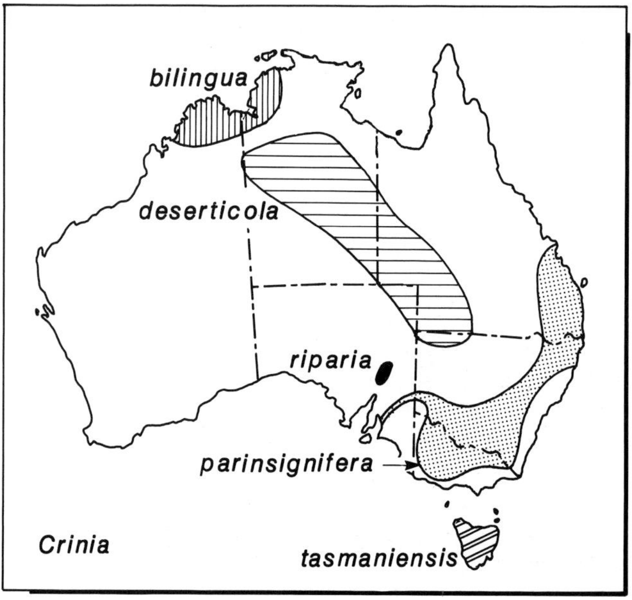

Crinia deserticola

Liem and Ingram

Desert Froglet

This species exhibits less variability than most. All individuals that we have seen are pale grey and bear the complex triangular and rectangular markings along the middle of the back. *Crinia deserticola* is one of the smallest members of the genus.

SIZE: Males 13–18 mm; females 15–20 mm.

HABITAT: Associated with static temporary or permanent water bodies where it shelters under leaves, corrugated iron, timber etc. Commonly found at the edge of turkey-nest dams in areas of blacksoil.

ADVERTISEMENT CALL: Resembles the chirruping sound of a sparrow.

BREEDING AND LARVAE: Spawn clumps are attached to submerged vegetation in swamps or pools. Tadpoles are a uniform, intense dark brown. The tadpole tooth formula is $\frac{1\ ^{1}\ 1}{1\ _{2}\ 1}$.

rinia deserticola, Townsville, Queensland *(Photo by Steve Richards).*

rinia georgiana, Walpole, Western Australia *(Photo by Gordon Grigg) (Specimen courtesy Grahame Webb).*

SIMILAR SPECIES: *Crinia deserticola* has a broad geographic range in arid and tropical areas. In the desert regions, small species are a rarity. There is a broad area of overlap with *C. bilingua* and *C. remota* and, in northern Queensland, with *C. deserticola*. Comparison of advertisement calls provides the only means of identification. *Crinia deserticola* might be confused with other small frogs such as *Uperoleia capitulata*, *U. rugosa* and *U. trachyderma*, but each of them can be distinguished from it by the possession of enlarged outer metatarsal tubercles.

Crinia georgiana

Tschudi

Tschudi's Froglet

Squat and flattened with a large head and short limbs. The dorsal skin can be smooth or tubercular and the abdomen is finely granular. The dorsal surfaces are grey, brown or almost black with or without elaborate patterns of lighter or darker longitudinal shades. The groin and the entire anterior and posterior surfaces of the thigh are bright red. The upper eyelids are red or golden.

SIZE: Males 24–32 mm; females 30–36 mm.

HABITAT: Areas of the coastal plain and forests where water collects in winter. Lives amongst vegetation beside creeks or soaks.

ADVERTISEMENT CALL: Described as "quack . . . quack. . . quack".

BREEDING AND LARVAE: Breeding takes place from July to October. About 70 large eggs are laid separately in shallow water. Tadpoles have a golden sheen, whilst the tail is largely transparent except for scattered white patches. Metamorphosis is completed in four to seven weeks. The tadpole tooth formula is $\frac{1\ ^{2}\ 1}{1\ _{2}\ 1}$.

SIMILAR SPECIES: The brilliant red of the groin and thighs is not shared by any other species, thus there is unlikely to be confusion.

Crinia glauerti

Loveridge

Glauert's Froglet

This is a highly variable species, with grey, brown, black or white markings. Females and some males have extensive areas of black on the undersurface.

SIZE: Males 20–22 mm; females 22–24 mm.

HABITAT: Areas of permanent moisture such as at the edge of marshes.

ADVERTISEMENT CALL: A long, hollow rattle resembling the noise of a pea being shaken in a can.

BREEDING AND LARVAE: Approximately 70 eggs are laid singly in shallow water, often associated with *C. georgiana*. Tadpoles are dark brown or blackish and take four months to complete metamorphosis.

SIMILAR SPECIES: Closely resembles *C. insignifera*, *C. pseudinsignifera* and *C. subinsignifera* and is distinguished principally by the advertisement call (see above).

Crinia insignifera

Moore

Sign-bearing Froglet

This is a small species with a slender body and long, slender limbs. The skin is smooth, warty or bears skin folds. Coloration varies from grey to brown, in a diversity of patterns. There is often a dark, triangular mark between the eyes.

SIZE: Males 14–25 mm; females 16–29 mm.

HABITAT: Temporary swamps on the coastal plains between Gingin and Busselton. Also present on Rottnest Island.

ADVERTISEMENT CALL: *Crinia insignifera* is unique amongst other southwestern members of the genus in having populations characterized by different calls. The most common of these is likened to "cheep . . . cheep . . . cheep". *C. insignifera* calls during winter.

Crinia glauerti, 100 km south-east of Perth, Western Australia *(Photo by Da Roberts).*

Crinia insignifera, south-west Western Australia *(Photo by Gordon Grigg (Specimen courtesy Grahame Webb).*

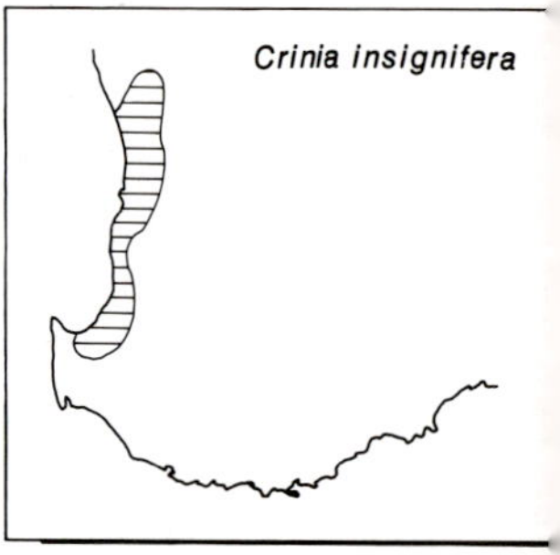

BREEDING AND LARVAE: Eggs are laid singly or in small clumps in shallow water and fall to the floor. Egg complements per female range from 66 to 268. The tadpoles take three to five months to complete metamorphosis.

SIMILAR SPECIES: Physically, *C. insignifera* resembles *C. pseudinsignifera* and *C. subinsignifera* very closely. Both are distinguished by their calls: *C. pseudinsignifera*, a wavering "baa . . . baa . . . baa", and *C. subinsignifera*, a long, low-pitched squelch.

Crinia parinsignifera

Main

Eastern Sign-bearing Froglet

The colour patterns and skin texture of this species are highly variable. The snout is semi-circular in profile and the ventral surface of the body is dark grey with a few scattered white spots on the abdomen, particularly in females.

SIZE: Males 18–22 mm; females 21–23 mm.

HABITAT: Found beneath debris at the edge of swamps and ponds.

ADVERTISEMENT CALL: Likened to the noise produced by drawing a wet finger over an inflated balloon (Tape track 40).

BREEDING AND LARVAE: Breeding takes place principally in mid-winter. Metamorphosis is completed in 11–12 weeks. The tadpole tooth formula is $\frac{1\,^{1}\,1}{1\,_{2}\,1}$.

SIMILAR SPECIES: Most likely to be confused with *C. signifera* and distinguished reliably only by comparison of the advertisement calls.

Crinia parinsignifera, Chowilla, South Australia *(Photo by Mike Mahony).*

Crinia pseudinsignifera, Darling Range, Western Australia *(Photo by Stev Donnellan).*

Crinia pseudinsignifera

Main

False Western Froglet

In its appearance this species closely resembles *C. insignifera*.

SIZE: Males 14–23 mm; females 16–29 mm.

HABITAT: Restricted to temporary swamps associated with run-off from granites on the Pre-Cambrian shield on the extreme west of Western Australia.

ADVERTISEMENT CALL: Three or four pulses in a high-pitched, wavering "baa, baa, baa". Males call in winter from July to September.

BREEDING AND LARVAE:From 77 to 207 eggs are laid singly in shallow water. Metamorphosis is copleted after 82–130 days. Tadpoles are mottled brown and the tadpole tooth formula is $\frac{1 \; ^{1} \; 1}{2}$.

SIMILAR SPECIES: Morphologically *Crinia pseudinsignifera* is similar to *C. insignifera*, *C. subinsignifera* and *C. glauerti*. It is distinguished from each of these species by its advertisement call (see above).

Crinia remota

Tyler and Parker

Remote Froglet

This species was described from southern New Guinea at a time when nothing was known about the *Crinia* species of northern Australia, but it is now known that *C. remota* occurs in north Queensland also and in the Northern Territory.

Crinia remota is one of the smallest frog species in Australia. The dorsum may be smooth or warty and the colour patterns highly variable, including lyrate or non-lyrate markings.

SIZE: Males 13–16 mm; females 14–19 mm.

HABITAT: Occurs in diverse areas such as flooded grassland or at the edge of swamps and billabongs.

Crinia remota, Indian Head Falls, Cape York, Queensland *(Photo by Glen Hoye).*

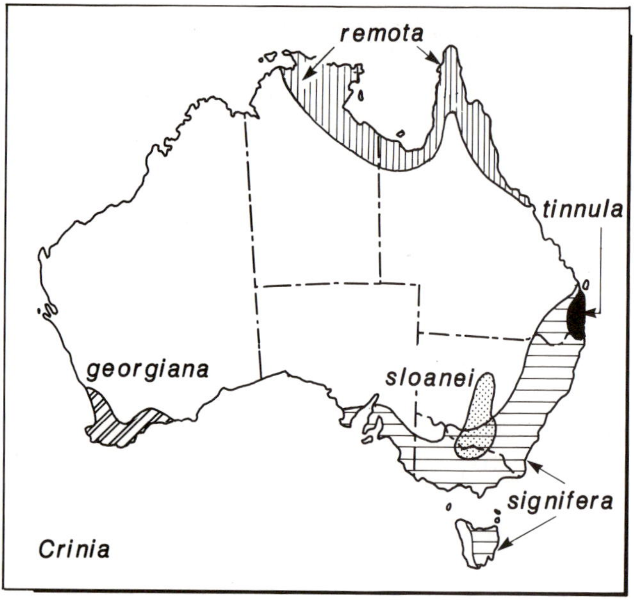

ADVERTISEMENT CALL: A high-pitched buzz comprising 12–15 pulses each with a duration of 720 msec.

BREEDING AND LARVAE: Unknown.

SIMILAR SPECIES: Most similar to *C. bilingua* (see p. 284).

Crinia riparia

Littlejohn and Martin

Streambank Froglet

Crinia riparia is the only *Crinia* species whose tadpoles are adapted to fast-flowing creeks. Also, it is unique within the genus in lacking a tympanum. In other respects it exhibits the variability of colour pattern and skin texture characteristic of the genus.

SIZE: Males 16–22 mm; females 19–25 mm.

HABITAT: Found beneath boulders and stones on the edge of rock-strewn creeks in the Flinders Ranges of South Australia.

ADVERTISEMENT CALL: A soft "kra-a-a-a-a-ack" repeated slowly.

BREEDING AND LARVAE: Spawn is laid beneath rocks at the edge of creeks. The tadpole has a flattened body and poorly developed tail fins. The tadpole tooth formula is $\frac{2}{1\;\;1 \atop 2}$.

SIMILAR SPECIES: The only species likely to be confused with it is *C. signifera* which replaces *C. riparia* at the southern end of the Flinders Ranges. In addition to the difference in call, *C. signifera* has a tympanum (absent in *C. riparia*).

Crinia signifera

Girard

Common Froglet

This is one of the most variable species, with a great diversity of colour and skin textures occurring in a single population. The tympanum is present. The fingers and toes are cylindrical and have lateral fringes but no webbing. The skin of the undersurface is granular.

Crinia riparia, Flinders Ranges, South Australia *(Photo by Mike Mahony).*

Crinia signifera, Watagan State Forest, New South Wales *(Photo by Mike Mahony).*

SIZE: Males 18–25 mm; females 19–28 mm.

HABITAT: Found beneath rocks, vegetation and debris at the edge of creeks, ponds, swamps and areas of seepage.

ADVERTISEMENT CALL: A rapid, repeated "crick . . . crick . . . crick . . . crick . . ." (Tape track 8).

BREEDING AND LARVAE: Breeds throughout the year except in mid-summer. Approximately 100–150 eggs are laid in small clumps in shallow water. Metamorphosis has taken seven weeks at 14–15°C. The tadpole tooth formula is $\frac{1\;^{1}\;1}{1\;_{2}\;1}$.

SIMILAR SPECIES: Most readily distinguished from other *Crinia* species by its call.

Crinia sloanei

Littlejohn

Sloane's Froglet

Crinia sloanei is exceptional amongst *Crinia* species in having only a single colour pattern: the back is always a mustard yellow or greyish with brighter yellow lyrate ridges upon the shoulders and large patches of darker pigment over the entire body. The throat of males is greyish green.

SIZE: Males 15–16 mm; females 16–17 mm.

HABITAT: Closely associated with temporary ponds.

ADVERTISEMENT CALL: Described as a short metallic "chick . . . chick . . . chick . . . chick" repeated frequently at a rate of one per second. Males call whilst floating in water.

BREEDING AND LARVAE: Unknown.

SIMILAR SPECIES: *Crinia parinsignifera* and *C. signifera* have distributions that are sufficiently close to that of *C. sloanei* to create possible confusion. Despite the consistent coloration of *C. sloanei*, call comparisons remain the best means of identification.

Crinia sloanei, Nyngan, New South Wales *(Photo by Harold Cogger).*

Crinia subinsignifera, Northcliffe, Western Australia *(Photo by Harold Cogger).*

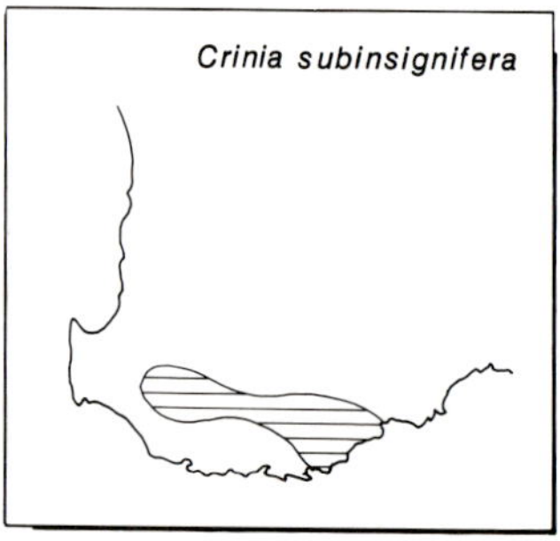

Crinia subinsignifera

Littlejohn

Small Western Froglet

In its appearance this species closely resembles *C. insignifera*.

SIZE: Males 14–20 mm; females 20–24 mm.

HABITAT: Temporary *Melaleuca* swamps in the extreme south-west of Western Australia from the vicinity of Manjimup south and east to Cheyne Beach.

ADVERTISEMENT CALL: Described as a long low-pitched squelch.

BREEDING AND LARVAE: Unknown.

SIMILAR SPECIES: *Crinia insignifera, C. glauerti* and *C. pseudinsignifera* are largely indistinguishable from this species, and correct identification requires comparison of calls (see the description for each species).

Crinia tasmaniensis

Günther

Tasmanian Froglet

Most individuals are grey or pale brown with longitudinally arranged patterns of darker pigment. The skin is smooth, warty or with raised ridges. Sometimes the tips of the warts are tinged with orange. The undersurface of the body and limbs is variegated with white or red on black.

SIZE: Males 21–23 mm; females 23–29 mm.

HABITAT: Always found near water, most commonly at the edge of streams or roadside gutters, but also near marshes or areas of shallow seepage.

ADVERTISEMENT CALL: Described as a quavering "bleat" repeated slowly and at irregular intervals. (Tape track 34).

BREEDING AND LARVAE: Eggs are laid in spring, singly or in small clumps. Tadpoles are uniform dark brown above, attain a total length of 25 mm, and complete their development in three to four months.

Crinia tasmaniensis, Mt Wellington, Tasmania *(Photo by Mike Mahony)*.

Crinia tasmaniensis, ventral view, Mt Wellington, Tasmania *(Photo by Mike Mahony)*.

SIMILAR SPECIES: Confined to Tasmania, the red markings on the undersurface will distinguish *C. tasmaniensis* from other *Crinia* species, but might create confusion with *Geocrinia laevis*. The latter lacks a fold of skin on the tarsus and fringes on the toes: features of *Crinia* species.

Crinia tinnula

Straughan and Main

Tinkling Froglet

Although the dorsal pattern of *C. tinnula* includes the range of colour patterns typical of the genus, *C. tinnula* is distinctive: the shape of the snout in profile projects beyond the lower jaw. In addition, the undersurface is a uniform dull grey with a faint, straight row of dots down the midline of the throat. This is a particularly small species.

SIZE: Males 16–18 mm; females 16–18 mm.

HABITAT: South-east coastal Queensland to north-east New South Wales in acid *Melaleuca* swamps, and wallum areas with poor drainage.

ADVERTISEMENT CALL: Described as "a short high-pitched ringing 'tching . . . tching' " likened to the sound of a small bell, made whilst the male is within sedge tussocks or at the water level.

BREEDING AND LARVAE: Amplexus and spawning has been recorded in July and August. The pH of the water in which spawn was deposited ranged from 4.3 to 5.2. Egg complements ranged from 33 to 118. The form of the tadpole has not been reported.

SIMILAR SPECIES: The geographic range of *C. tinnula* overlaps that of *C. parinsignifera* and *C. signifera*. *Crinia tinnula* can be distinguished by the features of ventral coloration, snout shape and size and advertisement call.

Crinia tinnula, Rainbow Beach, Queensland *(Photo by Mike Mahony)*.

Genus *Geocrinia* Blake

The genus *Geocrinia* was established by A. J. D. Blake in 1973 for a group of small frogs formerly referred to *Crinia*. Blake argued that the group was so distinctive that it should be considered a separate unit. What the *Geocrinia* species share when compared with *Crinia* is a slightly sturdier body, a smooth skin upon the undersurface of the body (granular in *Crinia* species) and, most importantly, a different reproductive habit: laying eggs out of instead of in the water. In some species of *Geocrinia* the entire period of development is spent within the egg capsule. In others the tadpoles develop initially within the egg membranes and are then released by flooding to complete the development in water.

Geocrinia currently includes seven species, five of which live in Western Australia. Two of these were described in 1991. The remaining two species are confined to south-east Australia.

Key to *Geocrinia* species

1a. Confined to southeastern Australia 2
b. Confined to Western Australia 3

2a. Call consists of a long introductory note, followed by a series of short musical notes; restricted to central and western Victoria, and the extreme south-east of New South Wales .... *G. victoriana* (p. 309)
b. Call consists of a harsh musical note repeated in a series; occurs in eastern and southeastern Victoria, southeastern South Australia and Tasmania. .... *G. laevis* (p. 305)

3a. Tips of toes dilated, abdomen greenish ..... *G. leai* (p. 305)
b. Tips of toes not dilated, abdomen not greenish 4

4a. Abdomen pink .... *G. rosea* (p. 309)
b. Abdomen not pink 5

5a. Abdomen bright yellow .... *G. vitellina* (p. 311)
b. Abdomen not bright yellow 6

6a. Abdomen white or pale yellowish .... *G. alba* (p. 303)
b. Abdomen fawn-brown .... *G. lutea* (p. 307)

Geocrinia alba

Wardell-Johnson and Roberts

White-bellied Frog

A small and squat species lacking expanded discs on the fingers and toes. The dorsal surface is light brown with large, darker brown spots and the abdomen is white or with a faint yellowish hue.

SIZE: Males 20–24 mm; female 17 mm.

HABITAT: Dense vegetation on clay soils.

ADVERTISEMENT CALL: A rapid series of 11–18 pulses.

BREEDING AND LARVAE: The spawn is laid in depressions in the soil. Details of development are unknown.

SIMILAR SPECIES: As indicated in the key, *G. alba* is distinguished from other *Geocrinia*, and certainly those also found in the extreme south-west of Western Australia, by the colour of the abdomen.

Geocrinia alba, Witchcliffe, Western Australia *(Photo by Dale Roberts).*

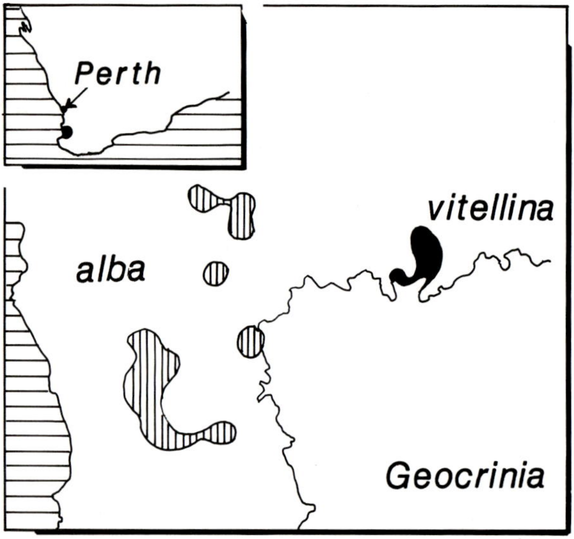

Geocrinia laevis

(Günther)

Smooth Frog

Physically this species is indistinguishable from *G. victoriana,* and so the preceding description applies equally well to *G. laevis*. The species is distinguished by its advertisement call.

SIZE: Males 23–27 mm; females 22–35 mm.

HABITAT: Dry sclerophyll and pine forests amongst leaf litter in areas subject to temporary flooding.

ADVERTISEMENT CALL: A variable number of pulsed notes, the first of which is the longest. Call likened to "cra-a-a-a-a-ack, cra-a-ack, cra-ack . . ."

BREEDING AND LARVAE: Littlejohn and Martin (1964) reported that large pigmented eggs are laid in March or April and deposited in loose, elongated masses that are often attached to vegetation. The number of eggs per mass is 76 to 147. The tadpoles are similar to those reported for *G. victoriana* and take approximately six months to complete development. The tadpole tooth formula is $\frac{1\ ^{1}\ 1}{1\ _{2}\ 1}$.

SIMILAR SPECIES: Very similar to *G. victoriana* but distinguished by call and except in western Victoria, not found in the same areas. Could be confused with *Pseudophryne* species and possibly even *Crinia signifera* but distinguished readily by the pale pink patches in the groin and on the undersurfaces of the legs.

Geocrinia leai

(Fletcher)

Lea's Frog

The overall colour is brown, with the front of the head, flanks and cloaca pale. The middle of the back and the legs are dark brown with small intense brown spots. The boundary of the two components may be separated by a narrow white line. The undersurfaces are pale green. The species is unusual amongst *Geocrinia* in having the tips of the toes dilated.

Geocrinia laevis, Tyrendarra, Victoria *(Photo by Murray Littlejohn).*

Geocrinia leai, Shannon Rock, Western Australia *(Photo by Mike Mahony).*

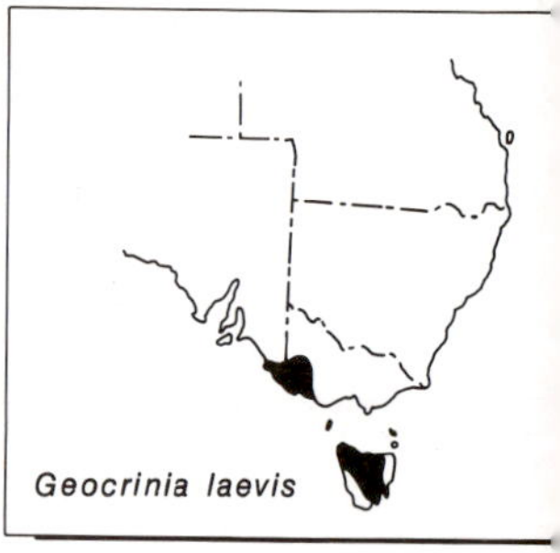

SIZE: Males 18–21 mm; females 23–26 mm.

HABITAT: Among vegetation near streams and swamps.

ADVERTISEMENT CALL: Likened to "chic . . . chic . . . chic . . . chic" by Main (1965).

BREEDING AND LARVAE: *Geocrinia leai* breeds in winter and is reported to lay spawn in clumps of 52–96 eggs attached to vegetation above water. The larvae fall into the water upon their emergence from the spawn clump. The larval span is approximately four months.

SIMILAR SPECIES: Overall body shape and adult size is shared by other *Geocrinia* species and by *Crinia* species, while the longitudinal arrangement of the dorsal markings bears a further resemblance to *Crinia*. By far the simplest means of recognizing *C. leai* is to examine the undersurface which is faint green, a colour not exhibited by other species.

Geocrinia lutea

(Main)

Walpole Frog

A small, squat bodied frog with a broad head and short and muscular limbs. The colour of the back is dark greyish brown with a darker mid-dorsal band. The throat is black in males and the abdomen pale fawn brown with faint red-brown blotches.

SIZE: Males 18–23 mm; females 22 mm.

HABITAT: Found in dense vegetation adjacent to streams around Walpole, Western Australia.

ADVERTISEMENT CALL: Reported to be similar to *G. rosea*, but repeated more rapidly.

BREEDING AND LARVAE: Males call from tunnels in clay or mud. Egg clumps of 25–30 are laid in the tunnels and take approximately two months to complete development. The tadpole does not feed.

SIMILAR SPECIES: Most similar to *G. rosea* (see p. 309).

Geocrinia lutea, Manjimup, Western Australia *(Photo by Steve Donnellan).*

Geocrinia rosea, Franklin River, Western Australia *(Photo by Steve Donnellan).*

Geocrinia rosea

(Harrison)

Karri Frog

A small species characterized by its short legs, smooth dorsal skin and with the back grey or brown bearing a broad dark brown longitudinal stripe down the centre. The abdomen is pink.

SIZE: Males 19–22 mm; females 24–25 mm.

HABITAT: Beneath dense vegetation, rotting logs, leaf litter, etc.

ADVERTISEMENT CALL: Described by Main (1965) as a series of short notes resembling "tk . . . tk . . . tk . . . tk".

BREEDING AND LARVAE: Main reports that from 26 to 32 eggs are laid out of water in shallow depressions or in tunnels in the Spring. The tadpoles remain mongst the broken-down egg membranes and complete their development there in 60–70 days.

SIMILAR SPECIES: In the geographic area where *G. rosea* is found, confusion is only likely with other *Geocrinia* species and possibly *Crinia*. From all of these species an examination of the colour of the undersurface of the body provides ready identification.

Geocrinia victoriana

(Boulenger)

Eastern Smooth Frog

This is a moderate sized species with a rounded snout and an indistinct tympanum. The limbs are short and the fingers and toes short and lack fringes. There is a small and indistinct inner but no outer metatarsal tubercule. The dorsal skin of the body and limbs is smooth and slightly roughened. The dorsal coloration overall is dull brown with obscure darker patches and, in some populations, occasional bright yellow, pale brown or reddish spots. The undersurface is densely stippled or mottled with dark brown, and there are patches of pink in the groin and on the tibia. The throat may be yellowish.

Geocrinia victoriana, Simpson's Creek, Victoria *(Photo by Gordon Grigg)*.

Geocrinia vitellina, Spearwood Creek, Western Australia *(Photo by Da[illegible] Roberts)*.

SIZE: Males 24–28 mm; females 21–33 mm.

HABITAT: Amongst low vegetation in sclerophyll forests extending to at least 1 500 m altitude.

ADVERTISEMENT CALL: The call has two components consisting of 1–3 long, pulsed notes resembling "cr-r-rack", followed by up to 50 shorter chirps resembling "pip,pip,pip,pip,pip,pip,pip . . .".

BREEDING AND LARVAE: Studies by Littlejohn and Martin (1964) revealed that 90–162 eggs are laid and that egg diameters range from 2.9 to 3.3 mm, the outer capsule being from 5.7 to 7.2 mm. In moist conditions the larvae can survive in the capsules for up to four months. The emerging tadpoles are 20–24 mm. The body is uniform dark brown and the fins lighter and speckled. Development is completed 6–8 months after hatching. The tooth formula is $\frac{1\ ^{1}\ 1}{3}$ with the outer of the lower rows particularly short.

SIMILAR SPECIES: *Geocrinia victoriana* overlap the distribution of other *Geocrinia* species only in a small area of central Victoria. In terms of size they might initially be mistaken for *Crinia signifera* which has particularly long toes (short in *G. victoriana*) and lacks the pink markings on the undersurface.

Geocrinia vitellina

Wardell-Johnson and Roberts

Orange-bellied Frog

A small, heavily built species with very short limbs and a roughened dorsal skin created by numerous warts. The fingers and toes are cylindrical and unwebbed. The back is pale brown or light grey with darker spots, whilst the undersurface is a brilliant orange-yellow (like egg-yolk).

SIZE: Males 21–24 mm; females 18 mm.

HABITAT: Confined to creek banks and areas of seepage.

ADVERTISEMENT CALL: A series of 9–15 pulses emitted slowly enough to be counted.

BREEDING AND LARVAE: Males call from depressions in clay beneath dense vegetation. The hatching tadpoles remain within the jelly of the spawn clump and do not feed. The tadpole has not been described.

SIMILAR SPECIES: The various *Geocrinia* species are distinguished from one another by differences in coloration. The only species sharing yellow on the undersurface of the body are *G. lutea* and *G. alba* which can be bright yellow to fawn but not the brilliant egg-yolk colour of *G. vitellina*.

Genus *Uperoleia* Gray

Of the 24 species of *Uperoleia* currently recognized, 18 have been described for the first time in the past 14 years. As a consequence little is known about the biology of most species and a study of *Uperoleia* is likely to be most rewarding for any herpetologist.

Uperoleia species are all small, squat frogs better known as "toadlets". They share a glandular skin with either discrete, raised glands, particularly at the back of the head or upon the flanks, or a rough tubercular skin leading to their common name association with toads. The limbs are always short, and the feet bear an inner metatarsal tubercle for burrowing.

To the ear the calls of *Uperoleia* species fall into two categories: the "clickers" and the "squelchers". The clickers make one or more distinct, identifiable clicks like stones being struck together. The "squelchers" have a call composed of numerous pulses which blur together into a "squelch".

It is rare for more than two species to be found together at any one locality.

Keys to *Uperoleia* species

Because *Uperoleia* species are so similar in their overall form and coloration, identification is very difficult. Species that are physically most similar to one another rarely exist together. Confirmation of identity by analysis of advertisement calls and skull anatomy is often required. Identification can be simplified by splitting the genus into geographic components. We therefore provide three keys.

Key to *Uperoleia* species of the Northern Territory

1a. Toes at least one-third webbed 2
b. Toes webbed at base or unwebbed 3

2a. Mid-vertebral stripe present .... *U. orientalis* (p. 337)
b. Lacking mid-vertebral stripe .... *U. borealis* (p. 318)

3a. Back with numerous, small, conical tubercles ....
.... *U. trachyderma* (p. 343)
b. Back lacking small, conical tubercles 4

4a. Distance between nostrils greater than distance from nostril to eye .... *U. micromeles* (p. 333)
b. Distance between nostrils less than distance from nostril to eye 5

5a. Parotoid and inguinal glands tipped with gold. Advertisement call a short click .... *U. lithomoda* (p. 328)
b. Parotoid and inguinal glands not tipped with gold. Advertisement call a long squelch 6

6a. Confined to Arnhem Land escarpment . *U. arenicola* (p. 315)
b. Broadly distributed upon the flood plains ....
.... *U. inundata* (p. 325)

Key to the *Uperoleia* species of Queensland, New South Wales, South Australia and Victoria

1a. Upper jaw with maxillary teeth 2
b. Upper jaw lacking maxillary teeth 6

2a. Ventral surface dark and heavily pigmented 3
b. Ventral surface pale with light pigmentation 5

3a. Parotoid glands enlarged and raised above surface of neck 4
b. Parotoid glands not enlarged .... *U. fusca* (p. 323)

4a. Back smooth, not flecked with light pigment ....
.... *U. tyleri* (p. 343)
b. Back rough and granular, flecked with light pigment ....
.... *U. martini* (p. 331)

5a. Distributed north of Townsville .... *U. altissima* (p. 315)
b. Distributed south of Townsville .... *U. laevigata* (p. 327)

6a. Back with numerous, fine, conical tubercles .... *U. trachyderma* (p. 343)
b. Back lacking fine, conical tubercles 7

7a. Parotoid and inguinal glands enlarged . *U. capitulata* (p. 320)
b. Parotoid and inguinal glands not enlarged 8

8a. Advertisement call a short click 9
b. Advertisement call a long squelch ..... *U. inundata* (p. 325)

9a. Back grey with well defined chocolate markings 10
b. Back with poorly defined markings 11

10a. Parotoid glands enlarged .... *U. littlejohni* (p. 330)
b. Parotoid glands not enlarged .... *U. rugosa* (p. 339)

11a. Parotoid and inguinal glands with gold flecks; advertisement call an unpulsed click .... *U. lithomoda* (p. 328)
b. Parotoid and inguinal glands lack gold flecks; advertisement call pulsed .... *U. mimula* (p. 334)

Key to the *Uperoleia* species of Western Australia

1a. Upper jaw toothed 2
b. Upper jaw lacking teeth 4

2a. Papilla on heel .... *U. mjobergii* (p. 337)
b. No papilla on heel 3

3a. Distance between nostrils greater than distance between eye and nostril; Tanami Desert .... *U. micromeles* (p. 333)
b. Distance between nostrils less than distance between eye and nostril; Western Kimberley .... *U. marmorata* (p. 330)

4a. Advertisement call a short click 5
b. Advertisement call a long squelch 9

5a. Distributed in the Kimberley 6
b. Confined to the Pilbara .... *U. glandulosa* (p. 325)

6a. Dorsum with faint, mid-dorsal stripe. Parotoid and inguinal glands tipped with gold .... *U. lithomoda* (p. 328)
b. Dorsum without faint, mid-dorsal stripe. Parotoid and inguinal glands not tipped with gold 7

7a. Dorsum covered with numerous, conical tubercles; many orange-tipped .... *U. trachyderma* (p. 343)
b. Dorsum not covered with numerous conical tubercles ... 8

8a. Call a fast click (about 3 pulses per second) . *U. aspera* (p. 318)
b. Call an unpulsed click .... *U. minima* (p. 336)

9a. Toes more than one-third webbed 10
b. Toes unwebbed or webbed at base .... *U. crassa* (p. 322)

10a. Back with distinct tubercles 11
b. Back smooth or with few tubercles .... *U. talpa* (p. 341)

11a. Parotoid glands vast; males 27–35 mm; confined to Pilbara .... *U. russelli* (p. 341)
b. Parotoid gland distinct; males 22–27 mm; confined to Kimberley .... *U. borealis* (p. 318)

Uperoleia altissima

Davies, Watson, McDonald, Trenerry and Werren

Montane Toadlet

Described in 1993, *U. altissima* is a small species confined to elevated areas on the Atherton and Windsor Tablelands in far northern Queensland. The diagnostic features are the possession of maxillary teeth, moderate development of the skin glands, grey dorsal colour with darker patches and a heavily pigmented undersurface. The groin and post-femoral patches are scarlet.

SIZE: Males 18–25 mm; females unknown.

HABITAT: Confined to moist eucalypt woodland and open forest.

ADVERTISEMENT CALL: A loud, sharp click.

BREEDING AND LARVAE: Unknown.

SIMILAR SPECIES: From the toothed species, *U. altissima* resembles *U. laevigata* and *U. fusca* closely in overall appearance, but is distinguished by having a short call, whereas those species have long calls of 32–68 and 20 pulses respectively.

Uperoleia arenicola

Tyler, Davies and Martin

Jabiru Toadlet

This small species lacks maxillary teeth. The skin ranges from smooth to weakly tubercular. The back is greyish, sandy or dull brown and the glands pale bronze. The groin and back of the thighs are orange-red. The fingers are long, slender, unfringed and unwebbed, and the toes long, fringed and unwebbed. The parotoid glands are well-defined, and the inguinal and coccygeal glands poorly defined.

SIZE: Males 18–23 mm; females unknown.

HABITAT: Known only from sandy soils at the edge of temporary creeks.

Uperoleia altissima, Millstream Falls, north Queensland *(Photo by Gordon Grigg*
(Specimen courtesy Margaret Davies and Graeme Watson).

Uperoleia arenicola, near Jabiru, Northern Territory *(Photo by Dick Zweifel).*

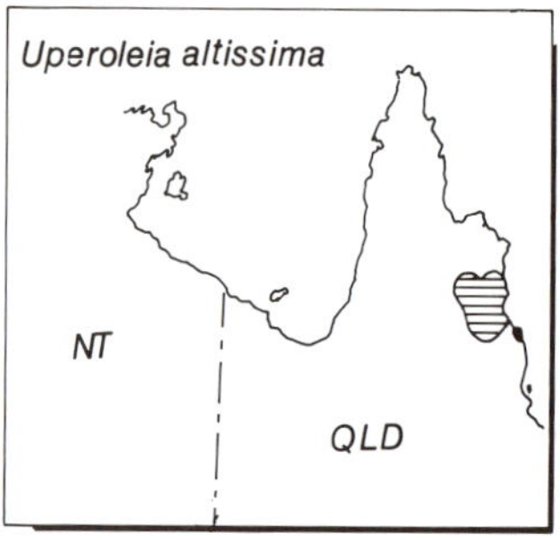

ADVERTISEMENT CALL: A short rasping call.

BREEDING AND LARVAE: Unknown.

SIMILAR SPECIES: Very closely related to *U. inundata* but distinguished by skeletal characters and by the pulse repetition rate and dominant frequency of the call. At present it appears that *U. arenicola* is confined to the western edge of the Arnhem Land escarpment of the Northern Territory, whereas *U. inundata* extends across the flood plains.

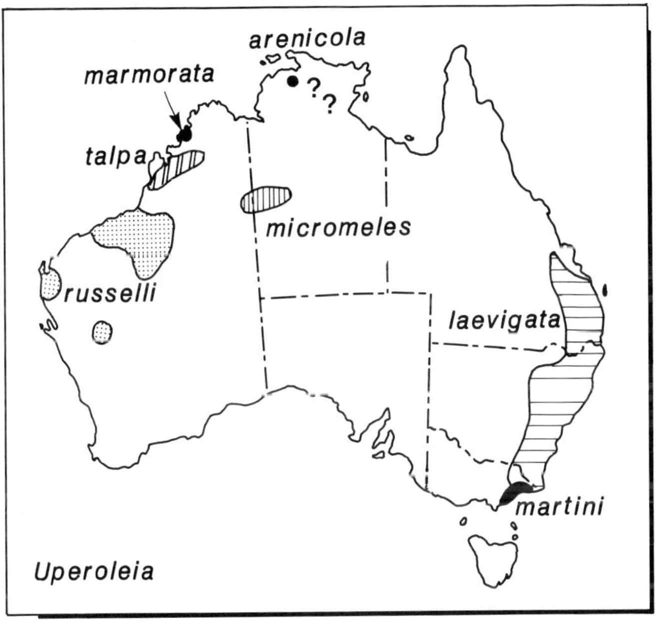

Uperoleia aspera

Tyler, Davies and Martin

Derby Toadlet

A moderate-sized Western Australian species with a slightly roughened dorsal skin mottled with small patches of dark brown upon a lighter brown background. The top of the head and the dermal glands are pale yellow masked with brown.

SIZE: Males 24–30 mm; females 29–34 mm.

HABITAT: Known only from grassland at the edge of filled gravel-scrapes. Confined to the vicinity of Derby and Broome in northern Western Australia.

ADVERTISEMENT CALL: A sharp, loud click repeated about three times a second.

BREEDING AND LARVAE: Unknown.

SIMILAR SPECIES: Only *U. mjobergii* and *U. talpa* exist in the same geographic area as *U. aspera*. *Uperoleia mjobergii* is readily distinguished by its possession of a heel lappet whereas *U. talpa* is a larger species (males 26–40 mm; females 32–38 mm) and bronze with or without small black patches.

Uperoleia borealis

Tyler, Davies and Martin

Northern Toadlet

A small, dull brown or slate species with slightly webbed toes, and orange patches in the groin and behind the thighs.

SIZE: Males 22–27 mm; females 26 mm.

HABITAT: Sparsely vegetated areas of temporarily inundated grassland.

ADVERTISEMENT CALL: A short rasping note. Males call from the base of grass tussocks or among stones in static or flowing water.

Uperoleia aspera, Derby, Western Australia *(Photo by Mike Mahony).*

Uperoleia borealis, Lake Argyle, Western Australia *(Photo by Margaret Davies).*

BREEDING AND LARVAE: Breeding has been observed in February. Eggs laid by a captive female had a capsule diameter of 1.75–1.96 mm.

SIMILAR SPECIES: Only *U. lithomoda* and *U. trachyderma* exist within the geographic range of *U. borealis. Uperoleia lithomoda* has an advertisement call consisting of sharp clicks, and *U. trachyderma* can be distinguished readily by its very rough dorsal skin.

Uperoleia capitulata

Davies, McDonald and Corben

Small-headed Toadlet

A very rough-skinned species of eastern Australia characterized by a small head, smooth skin with enlarged glands, and irregular markings upon the back.

SIZE: Males 19–27 mm; females 27–28 mm.

HABITAT: Within its extensive geographic range it has been recorded in mulga woodland and at waterholes bounded by coolibah.

ADVERTISEMENT CALL: Unknown.

BREEDING AND LARVAE: Unknown.

SIMILAR SPECIES: The geographic distribution of *U. capitulata* overlaps only with *U. rugosa*. The latter may be distinguished by its less conspicuous skin glands.

Uperoleia capitulata, Charters Towers, Queensland *(Photo by Margaret Davies)*.

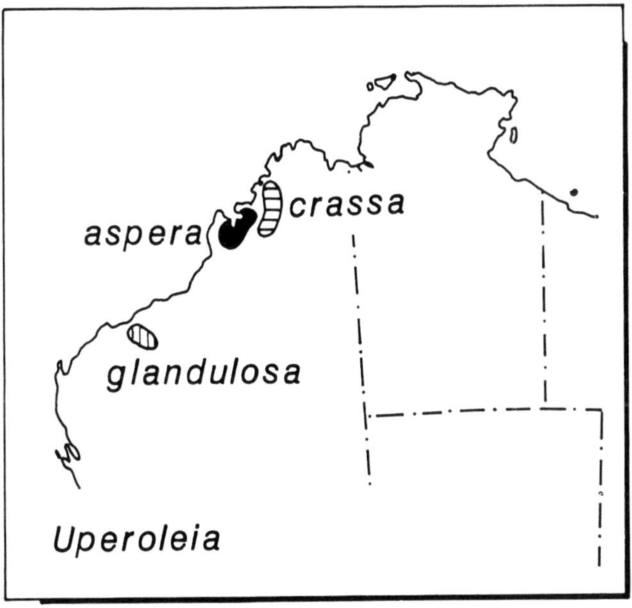
crassa
aspera
glandulosa
Uperoleia

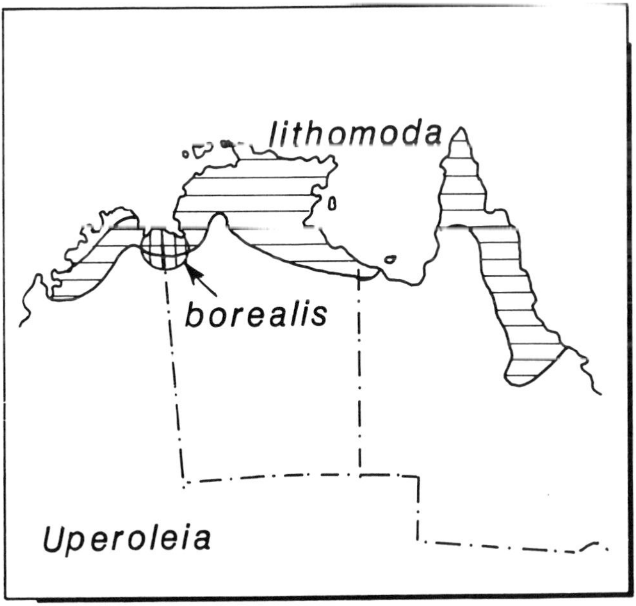
lithomoda
borealis
Uperoleia

Uperoleia crassa

Tyler, Davies and Martin

Fat Toadlet

A small rotund species known from the Mitchell Plateau in the remote western coast of the Kimberley Division of Western Australia. The back is uniform dark brown except for the parotoid glands which are golden. The sides of the head and the flanks are slate.

SIZE: Males 22–30 mm; females 30–31 mm.

HABITAT: Known only from partially flooded, cleared areas surrounded by sparse *Livistona* and *Eucalyptus* forest.

ADVERTISEMENT CALL: Males call from the base of grass tussocks adjacent to the water. The call is a brief, rasping sound.

BREEDING AND LARVAE: Breeding was observed in January. A total of 368 eggs were laid by a single individual. The capsule diameter ranged from 1.8 to 2.2 mm.

SIMILAR SPECIES: The only species collected near *U. crassa* is *U. minima* which is a distinctly smaller species (males 16–21 mm) and has a call in the form of a single click (not a rasping sound).

Uperoleia crassa, Mitchell Plateau, Western Australia *(Photo by Mike Mahony)*.

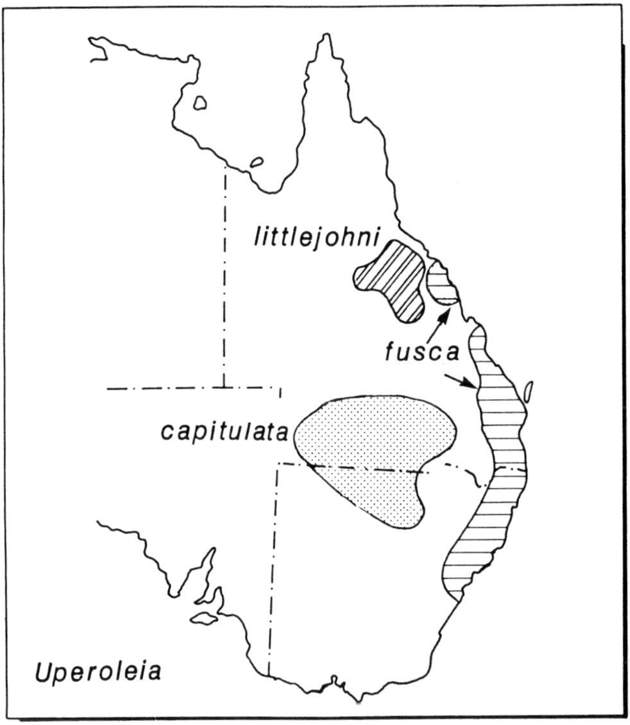

Uperoleia fusca

Davies, McDonald and Corben

Dusky Toadlet

This species is of moderate size, has a smooth or faintly rugose skin, maxillary teeth, moderately developed skin glands, and a slate to chocolate dorsal coloration. The undersurface is extensively marked with dark colours.

SIZE: Males 20–28 mm; females 23–29 mm.

HABITAT: Found most commonly in open forest with a tussock grass understorey.

ADVERTISEMENT CALL: A short rasping note lasting about one-third of a second.

BREEDING AND LARVAE: Unknown.

SIMILAR SPECIES: Of the toothed species of *Uperoleia* the most similar is probably *U. laevigata* from which it may be distinguished by the darkly pigmented ventral surface (unpigmented or very lightly pigmented in *U. laevigata*).

Uperoleia fusca, Eungella, Queensland *(Photo by Margaret Davies).*

Uperoleia glandulosa, Port Headland, Western Australia *(Photo by Steve Donnellan).*

Uperoleia glandulosa

Davies, Mahony and Roberts

Glandular Toadlet

A moderately sized species with prominent glands, fringed toes and a distinctive coloration which is predominantly olive brown with discrete black spots and blotches. There is a disrupted orange stripe extending from behind the eye to the groin, a small orange spot on each side of the sacrum and a narrow, orange, mid-vertebral stripe. The ventral surface is white with some small grey spots. The groin and post-femoral patches are orange red.

SIZE: Males 20–25 mm; females 25 mm.

HABITAT: Known from temporarily flooded areas such as creek banks and at the edge of claypans.

ADVERTISEMENT CALL: A single note composed of two or three pulses but, to the ear, a sharp click.

BREEDING AND LARVAE: Unknown.

SIMILAR SPECIES: The only species known to exist in the same area as *U. glandulosa* is *U. russelli* which has a long call and extensive webbing between the toes.

Uperoleia inundata

Tyler, Davies and Martin

Flood Plain Toadlet

Confined to the flood plains of the Northern Territory and north-west Queensland, this species lacks unique distinguishing characteristics and is recognized principally by its call. It lacks webbing between the toes and has a dark brown coloration with faint darker markings.

SIZE: Males 23–28 mm; females 24–30 mm.

HABITAT: Grassland and open forest at edge of flood plains.

ADVERTISEMENT CALL: A short, rasping call.

Uperoleia inundata, Kapalga, Northern Territory *(Photo by John Wombey).*

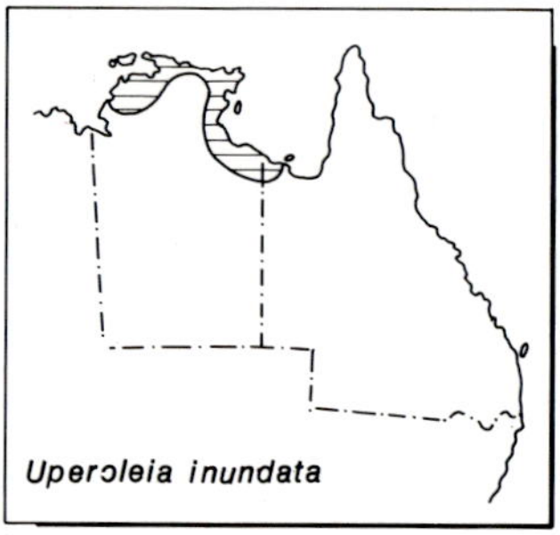

BREEDING AND LARVAE: Breeding takes place during January and February. Clumps of five or so eggs are laid together, or eggs are laid individually in very shallow water and sink to the bottom. The capsule diameter is 1.5–2.0 mm. Tadpoles have a short body and long tail and are dark brown and attain a maximum length of 34 mm. Larval development time is uncertain; eight weeks has been reported but shorter periods are likely. The tooth formula is $\frac{1\;^{1}\;1}{3}$ but it should be noted that the lower row is confined to a remnant in the midline.

SIMILAR SPECIES: *Uperoleia arenicola* and *U. lithomoda* are the only species to be found on the flood plains of the Northern Territory with *U. inundata*. *Uperoleia arenicola* is best distinguished by its rasping call, with the individual notes repeated so rapidly that the call is blurred. In contrast *U. inundata* has a call in which the individual pulses can be recognized. *Uperoleia lithomoda* has a call that is a distinct, single click rather than a longer, rasping sound.

Uperoleia laevigata

Keferstein

Smooth Toadlet

This is one of the few species with maxillary teeth. The parotoid glands are generally extremely prominent and the fingers and toes are fringed. Almost all individuals have a triangular patch on the head. The ventral surface is pale with sparse darker stippling.

SIZE: Males 20–28 mm; females 22–32 mm.

HABITAT: Extends over a wide range of plant associations ranging from closed forest to grassland.

ADVERTISEMENT CALL: A long pulsed call extending for more than one half of a second. (Tape track 23).

BREEDING AND LARVAE: The tadpole of this species has a tooth formula $\frac{1}{3}$.

Uperoleia laevigata, Wollombi, New South Wales *(Photo by Gordon Grigg)*.

SIMILAR SPECIES: In the area in which *U. laevigata* is found there are only two other toothed species, *U. fusca* and *U. tyleri*. The former is distinguished most readily by the ventral pigmentation, which is dense in *U. fusca* and sparse in *U. laevigata*. The parotoid glands are much larger and more raised in *U. tyleri* than in *U. laevigata*.

Uperoleia lithomoda

Tyler, Davies and Martin

Stonemason Toadlet

A small dull brown species with unwebbed toes, a fine cream line bisecting the head and, commonly, gold patches or a gold line upon the sides of the body. The skin glands are prominent.

SIZE: Males 17–27 mm; females 24–29 mm.

HABITAT: Most commonly encountered in open grassland, including country that has been cleared.

ADVERTISEMENT CALL: A sharp click resembling the sound of two smooth stones being struck together. Males call from hidden or exposed situations near temporary water.

BREEDING AND LARVAE: Tadpoles are dark brown and have flattened bodies. The mouth disc is very small and the tooth formula is $\frac{1\ ^{1}\ 1}{3}$.

SIMILAR SPECIES: Because of its wide geographic range across northern Australia, *U. lithomoda* is associated with many other *Uperoleia* species at different parts of its geographic range. Its closest relative is *U. mimula* which in Australia is confined to the Cape York Peninsula of northern Queensland, and to the Torres Strait Islands. Both have "click" calls but in *U. mimula* more than one pulse can be detected. In addition *U. mimula* has bolder, dark markings.

Iperoleia lithomoda, Collinsville, Queensland *(Photo by Gordon Grigg) (Specimen courtesy Margaret Davies and Graeme Watson).*

Iperoleia littlejohni, Burra Range, Queensland *(Photo by Gordon Grigg) (Specimen courtesy Margaret Davies and Graeme Watson).*

Uperoleia littlejohni

Davies, McDonald and Corben

Littlejohn's Toadlet

Uperoleia littlejohni is a species confined to subcoastal northeastern and north central Queensland. It lacks teeth and toe webbing. The body is dull brown with chocolate patches and faint sandy markings upon the prominent parotoid glands, whilst the groin and thigh patches are scarlet.

SIZE: Males 20–32 mm; females 23–29 mm.

HABITAT: Found predominantly in open woodland dominated by Iron Bark and Bloodwood, males calling from the base of *Triodia* tussocks.

ADVERTISEMENT CALL: A loud, sharp click of detectable pulses.

BREEDING AND LARVAE: Unknown.

SIMILAR SPECIES: Distinguished from other species lacking teeth and toe webbing by the absence of heavy pigmentation upon the undersurface of the body, and by possession of a distinct pattern of marking upon the back.

Uperoleia marmorata

Gray

Marbled Toadlet

Described in 1841 from material collected on the first scientific expedition to the Kimberley, the genus *Uperoleia* was erected for this species. Despite numerous reported sightings of the species, it has not been recorded since the original description: all reports are of other species made at a time when it was not realized how many species exist in Australia. The original specimen is a male possessing teeth and lacking webbing between the toes. Gray stated that it was "black and green marbled, leaving a triangular greenish spot on the forehead, beneath lead colour".

Good *Uperoleia* habitat, Derby, Western Australia. Here were found *U. talpa*, *U. aspersa* and *U. mjobergi* (*Photo by Margaret Davies*).

SIZE: Male 30 mm; female unknown.

HABITAT: Thought to be an area south of the Prince Regent River, Kimberley Division, Western Australia. Details unknown.

ADVERTISEMENT CALL: Unknown.

BREEDING AND LARVAE: Unknown.

SIMILAR SPECIES: *Uperoleia mjobergii* is the only other toothed species of *Uperoleia* in the Kimberley, and the presence of a small tubercle on the heel of *U. mjobergii* will distinguish it from *U. marmorata*.

Uperoleia martini

Davies and Littlejohn

Martin's Toadlet

Uperoleia martini is confined to eastern, coastal Victoria and the adjacent portion of New South Wales. It is a large species which possesses teeth, enlarged parotoid glands, roughened dorsal skin, unwebbed toes and considerable dark pigmentation on the undersurface of the body. Known only from preserved specimens.

Uperoleia martini, Cann River, Victoria *(Photo by Gordon Grigg)*.

Uperoleia micromeles, Tanami Desert, Northern Territory *(Photo by Margaret Davies)*.

SIZE: Males 30–33 mm; females unknown.

HABITAT: Found in dry forest, woodlands, shrublands, grasslands, open and disturbed areas.

ADVERTISEMENT CALL: A highly pulsed call lasting for approximately 0.5 sec.

BREEDING AND LARVAE: Unknown.

SIMILAR SPECIES: The large size and possession of teeth are features shared in southeastern Australia only by *U. tyleri*. The calls of the species provide the clearest separation, the call of *U. tyleri* being about one-half of the duration of *U. martini*.

Uperoleia micromeles

Tyler, Davies and Martin

Tanami Toadlet

A robust species lacking teeth and webbing between the toes. Unique to this species is the extremely broadly spaced nostrils which are separated by a distance greater than the distance between the eye and the nostril. The back is pale brown flecked with cream and scarlet.

SIZE: Males 24–27 mm; females 24–31 mm.

HABITAT: Discovered upon a red soil sand plain bearing scattered mulga (*Acacia aneura*) and a dominant understorey of spinifex (*Triodia* species).

ADVERTISEMENT CALL: Unknown.

BREEDING AND LARVAE: Unknown.

SIMILAR SPECIES: *Uperoleia micromeles* is known only from the Tanami Desert in the Northern Territory, and the adjacent area of Western Australia. At present no other species has been reported from there, and *U. micromeles* is highly distinctive in having broadly spaced nostrils: the nostrils being separated by a distance greater than the distance between the nostril and the eye. In all other *Uperoleia* species the proportions are reversed.

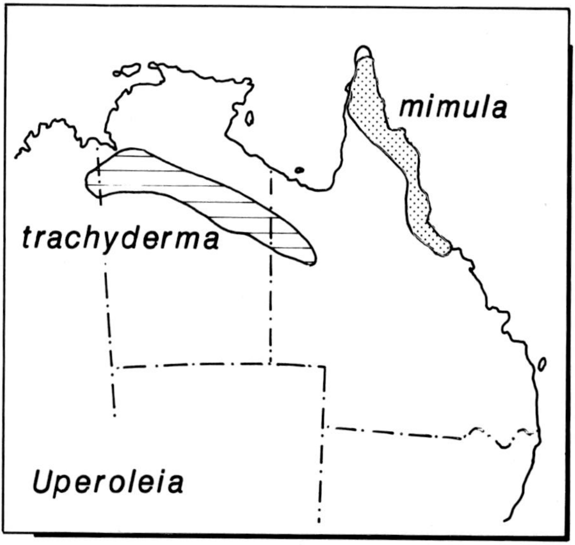

Uperoleia mimula

Davies, McDonald and Corben

Mimic Toadlet

A small to moderate-sized species confined to the eastern coast of the Cape York Peninsula of Queensland and as far south as Townsville. Maxillary teeth are absent and toe webbing basal or absent. The dorsal surface is dull brown with dark chocolate patches. The parotoid glands and flanks are faintly marked with a sandy colour. The undersurface is densely stippled with dull grey.

SIZE: Males 20–28 mm; females 21–28 mm.

HABITAT: *Melaleuca* woodland, eucalypt open woodland and woodland with tussock grassland.

ADVERTISEMENT CALL: A "click" of three to five pulses.

BREEDING AND LARVAE: Tadpoles grow to 20 mm, have an extremely small oral disc and the tooth formula the same as *U. inundata*.

SIMILAR SPECIES: The only species found within the geographic range of *U. mimula* is *U. lithomoda*. Both species have "click" calls, but that of *U. mimula* is distinctly pulsed, whereas that of *U. lithomoda* is not.

Uperoleia mimula, Townsville, north Queensland (Photo by Dick Zweifel).

Uperoleia minima, Mitchell Falls, Western Australia (*Photo by Steve Donnellan*).

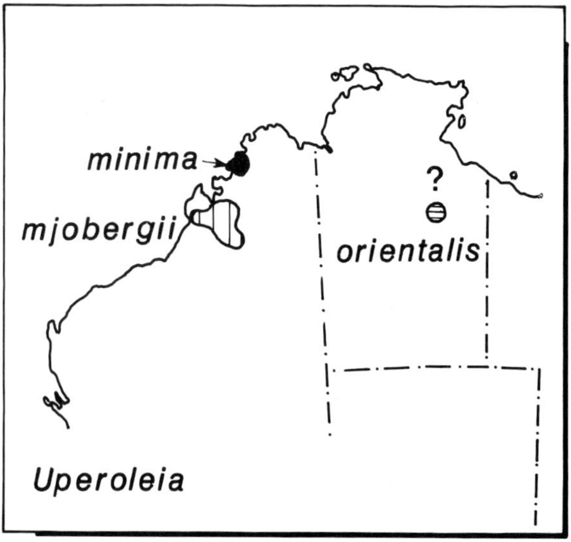

Uperoleia minima

Tyler, Davies and Martin

Small Toadlet

This tiny species is known only from a series of specimens collected on the Mitchell Plateau in the Kimberley Division of Western Australia. The toes are long and unwebbed and there are no teeth. The dorsal surface is roughened and dark brown with indistinct, irregular patches of darker pigment. Patches in the groin and behind the thighs are scarlet.

SIZE: Males 16–21 mm; females unknown.

HABITAT: Dense grassland.

ADVERTISEMENT CALL: A single, sharp click. Males call at the base of grass tussocks up to 20 m from the water.

BREEDING AND LARVAE: Unknown.

SIMILAR SPECIES: Of the Kimberley species approaching the small size of *U. minima*, *U. mjobergii* is readily distinguished by its heel tubercle, whilst *U. lithomoda* has golden markings on the sides, lacking in *U. minima*. *Uperoleia crassa* is found in the same general area but is larger and, also, has a rasping advertisement call.

Uperoleia mjobergii

(Andersson)

Mjoberg's Toadlet

This toothed species has a trace of webbing between the toes and a prominent tubercle on the heel. The back is dull brown with darker and lighter patches, and the parotoid glands are yellowish-brown.

SIZE: Males 19–25 mm; females 21–23 mm.

HABITAT: Open flood plains where it has been found by day sheltering under logs — other species burrow.

ADVERTISEMENT CALL: A short rasp or creak.

BREEDING AND LARVAE: Pairs have been observed in amplexus in February following cyclonic rains. Spawn and tadpoles unknown.

Uperoleia orientalis

(Parker)

Alexandria Toadlet

This enigmatic frog has been collected only once since its discovery in 1905. At least part of the problem is the fact that Alexandria Station, cited as the type locality, once occupied a substantial proportion of the Northern Territory. Hence, the exact collection site is unknown.

Uperoleia orientalis is distinctive in having moderately extensive webbing between the toes and a narrow, pale, mid-vertebral stripe.

SIZE: Males 26–28 mm; females unknown.

HABITAT: At Cox River, east of Nutwood Downs, a specimen was taken at a spring-fed swamp in paper-bark woodland.

ADVERTISEMENT CALL: Unknown.

BREEDING AND LARVAE: Unknown.

SIMILAR SPECIES: The large size and webbing between the toes is shared with *U. russelli* of the Kimberley Division of Western Australia, in which the webbing is more extensive. Further comparison will have to await the rediscovery of *U. orientalis*.

Uperoleia mjobergii, Derby, Western Australia *(Photo by Margaret Davies)*.

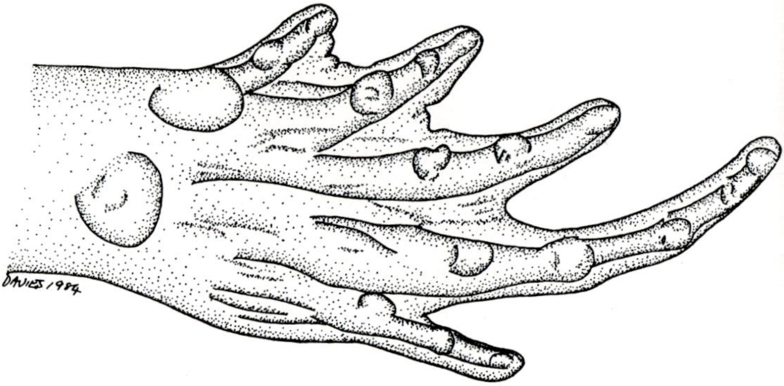

Foot of *Uperoleia orientalis* showing the extensive webbing between the toes. (From Transactions of the Royal Society of South Australia. Reproduced with permission.)

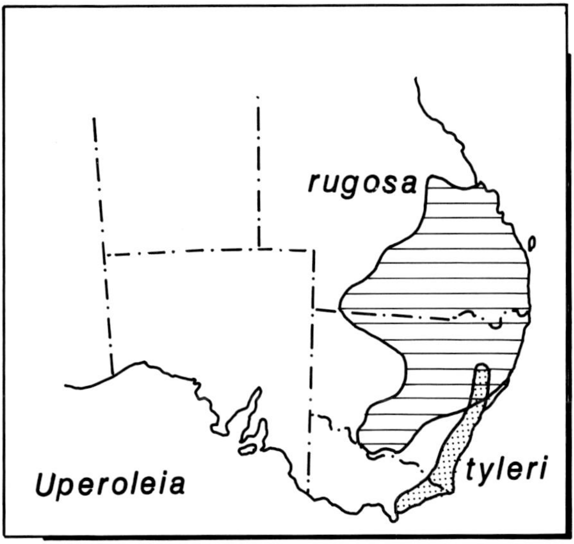

Uperoleia rugosa

(Andersson)

Wrinkled Toadlet

This species varies in size more than any other *Uperoleia* species, with small individuals in the coastal areas and larger ones inland. Teeth are absent. The toes are unfringed and may have basal webbing. The back is grey to brown with very distinct darker markings. Tubercles on the back are tipped with yellow, and the parotoid and inguinal glands are yellowish. The groin and thigh patches are reddish-orange.

SIZE: Males 18–32 mm; females 18–30 mm.

HABITAT: Forested slopes and grassy plains.

ADVERTISEMENT CALL: A high, distinctly pulsed click.

BREEDING AND LARVAE: Unknown.

SIMILAR SPECIES: *Uperoleia rugosa* coexists over a portion of its range with *U. capitulata*. The latter can be distinguished most readily by its smooth skin but larger skin glands.

Uperoleia rugosa, west of Dalby, Queensland *(Photo by Marion Anstis)*.

Uperoleia russelli, Port Headland, Western Australia *(Photo by Steve Donnellan)*.

Uperoleia russelli

(Loveridge)

Russell's Toadlet

Uperoleia russelli lacks teeth and is characterized by extremely extensive webbing between the toes, and by its vast skin glands. The back is pale brown with numerous, small, darker spots. The parotoid and inguinal glands are yellowish.

SIZE: Males 27–35 mm; females 30–35 mm.

HABITAT: Found around flooded claypans, beside roadside drainage channels and associated areas.

ADVERTISEMENT CALL: Distinctly pulsed and likened to "qrk . . . qrk . . . qrk". Said to be short and grating.

BREEDING AND LARVAE: Unknown.

SIMILAR SPECIES: *Uperoleia talpa* of the Kimberley is most similar in size, but has less webbing between the toes and a longer call.

Uperoleia talpa

Tyler, Davies and Martin

Mole Toadlet

Uperoleia talpa lacks teeth, has extensive webbing between the toes, and possesses a smooth skin and large glands. The back is pale bronze with a few irregular islands of darker pigment upon it. The submandibular gland is white.

SIZE: Males 26–40 mm; females 35–38 mm.

HABITAT: In areas of flooded grassland with sandy soils.

ADVERTISEMENT CALL: A relatively long creaking sound. Males call from leaf litter and from holes at the base of isolated eucalypt and boab trees.

BREEDING AND LARVAE: Spawn has been collected in February.

SIMILAR SPECIES: *Uperoleia russelli* is the only other species with extensive toe webbing. It differs in having contrasting skin glands of pale yellow, and a shorter call.

Uperoleia talpa, Gibb River, Western Australia *(Photo by Margaret Davies).*

Uperoleia trachyderma, Rankine River, Northern Territory *(Photo by Margaret Davies).*

Uperoleia trachyderma

Tyler, Davies and Martin

Blacksoil Toadlet

This species lacks teeth and has fringed toes. It is unique amongst *Uperoleia* species in being covered dorsally with small, conical tubercles. The back is light grey marked with extensive areas of deep orange on the glands and many of the tubercles are tipped in the same colour.

SIZE: Males 17–26 mm; females unknown.

HABITAT: Amongst grass on blacksoil plains.

ADVERTISEMENT CALL: A harsh creak of four pulses.

BREEDING AND LARVAE: Unknown.

SIMILAR SPECIES: The dorsal skin texture and coloration are so distinctive that this species should not be confused with any other.

Uperoleia tyleri

Davies and Littlejohn

Tyler's Toadlet

Uperoleia tyleri is a large, toothed species with unwebbed toes and enlarged parotoid glands. The back is a dark brown/black with orange-yellow spotting upon the skin glands. The groin and thigh markings are yellow, and the undersurface of the body is bluish-black with white spots.

SIZE: Males 22–23 mm; females 26–34 mm.

HABITAT: Found in dry forest, woodlands, shrublands, grasslands, open and disturbed areas.

ADVERTISEMENT CALL: A single, short, pulsed note.

BREEDING AND LARVAE: The eggs are small (1.5 mm diameter) and pigmented. The tadpole tooth formula is $\frac{1\,^{1}\,1}{1\,_{2}\,1}$.

SIMILAR SPECIES: Confusion may arise with *U. martini* and *U. laevigata*. The relatively dense ventral pigmentation serves to distinguish *U. tyleri* from these species.

Uperoleia tyleri, Upper Colo, New South Wales *(Photo by Gordon Grigg)*.

Genus *Lechriodus* Boulenger

Lechriodus is unique amongst myobatrachid frog genera in having more species in New Guinea than in Australia. The single Australian representative is *L. fletcheri* which is confined to a narrow zone south of the New South Wales/Queensland border. *Lechriodus* species live on the forest floor and in hollows of trees and are well camouflaged.

Lechriodus fletcheri

(Boulenger)

Fletcher's Frog

The head and body are depressed and the legs long and muscular. The fingers and toes are long and unwebbed. The dorsal skin is finely granular and there are narrow raised skin folds: from the posterior corners of the eye to above the insertion of the forearms, and a V-shaped pair between them. The colour ranges from dull yellow through various shades of brown to dull chocolate. There is a narrow and irregular black line from the nostril to the eye, continuing behind the eye along the path of the skin fold. The hind limbs are commonly marked with narrow transverse bars.

Lechriodus fletcheri, Wiangaree, New South Wales *(Photo by Mike Mahony)*.

Lechriodus fletcheri, pair in amplexus, Ourimbah, New South Wales *(Photo by Gordon Grigg)*.

SIZE: Males 42–48 mm; females 45–54 mm.

HABITAT: Amongst leaves and other groundcover on the forest floor, and in tree hollows.

ADVERTISEMENT CALL: A purring "g-r-r-r-r-up" lasting for about one second. (Tape track 20).

BREEDING AND LARVAE: Approximately 300 eggs are laid in a foam nest usually in a small body of water. The tadpoles are commonly cannibalistic and will also eat the eggs and tadpoles of other species. The tadpole tooth formula is $\frac{5\ ^{1}\ 5}{1\ _{2}\ 1}$.

SIMILAR SPECIES: *Lechriodus fletcheri* bears a resemblance to *Mixophyes* species in coloration, including the black stripe on the side of the head and transverse bars on the legs. The latter are larger and have distinct webbing between the toes.

Genus *Assa* Tyler

The sole member of the genus *Assa* lives in temperate rainforest on the border of southeastern Queensland and northeastern New South Wales. The species *darlingtoni* was described in 1933 by A. Loveridge who placed it in the genus *Crinia*. In 1966 I. R. Straughan and A. R. Main discovered that the male frog transports the tadpoles in a pair of "hip pockets". M. J. Tyler created the new genus *Assa* in 1972 for the species.

Assa darlingtoni

(Loveridge)

Hip-pocket Frog, Marsupial Frog

The body is broad and the arms, fingers and toes are very slender. The legs are short but muscular. The skin of the back is smooth or slightly roughened and finely warty on the legs.

The background coloration ranges from grey to brown, usually with distinctive darker patches upon the head and body and a narrow dark stripe upon the flanks.

Assa darlingtoni, Mt Warning, Queensland *(Photo by Gordon Grigg)*.

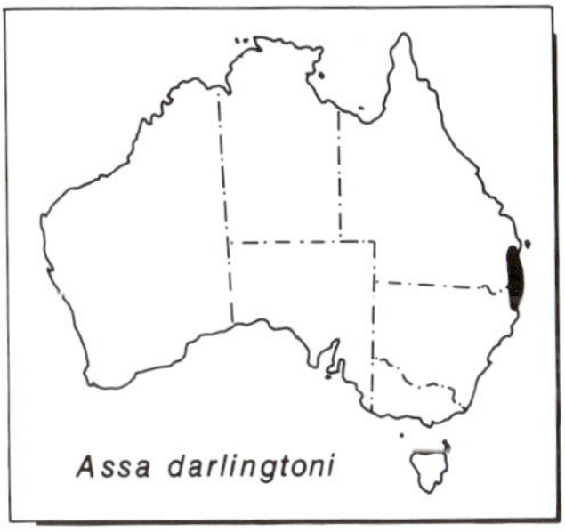

SIZE: Males 15–19 mm; females 18–21 mm.

HABITAT: Amongst leaf-litter in montane forest.

ADVERTISEMENT CALL: A series of 6–10 harsh notes rapidly repeated: "eh eh eh eh eh eh . . .". (Tape track 29).

BREEDING AND LARVAE: About 10 or 11 eggs are laid in late summer in a large mass of jelly upon the ground. The male approaches the hatching tadpoles and permits them to wriggle upon his body and into the hip pouches. They remain there for about two months before leaving the pouches as tiny frogs.

SIMILAR SPECIES: The only species living in the same area that might be confused with *A. darlingtoni* is *Kyarranus loveridgei*. The latter has broader arms and angular skin folds upon the back.

Genus *Adelotus* Ogilby

The single species of *Adelotus* differs from all other genera of Australian frogs in three respects: the male is larger than the female, the head of the male is consistently larger than the head of the female, whilst the species has sharp, tooth-like tusks that project from the lower jaw.

Adelotus brevis

(Günther)

Tusked Frog

The head of the male is often so large that it is equivalent to the size of the body. The dorsal surface is covered with small warts. The fingers and toes are cylindrical and are not webbed. The upper surfaces of the body are dull slate, whilst the head is pale brown. The lower surfaces are densely variegated with black upon white, whilst the groin and the calves bear flashes of red upon black.

SIZE: Males 34–44 mm; females 29–38 mm.

HABITAT: Found in open country or in forests.

ADVERTISEMENT CALL: A slowly repeated "chuluk", reminiscent of *Limnodynastes peronii* but softer and just detectably biphasic. (Tape track 21).

BREEDING AND LARVAE: The unpigmented eggs are laid in a foam nest usually hidden from the light. At 20°C development is completed in 71 days. The tadpole tooth formula is $\frac{2\ ^{1}\ 2}{1\ _{2}\ 1}$.

SIMILAR SPECIES: Small individuals might be confused with *Pseudophryne* species, particularly because of the dense pigmentation on the undersurface. However, the large head of *A. brevis* is noticeable even in juveniles, whilst the presence of the tusks will quickly distinguish it.

Adelotus brevis, Barrington House, New South Wales *(Photo by Gordon Grigg)*.

Adelotus brevis, ventral view, Gosford, New South Wales *(Photo by Gordon Grigg)*.

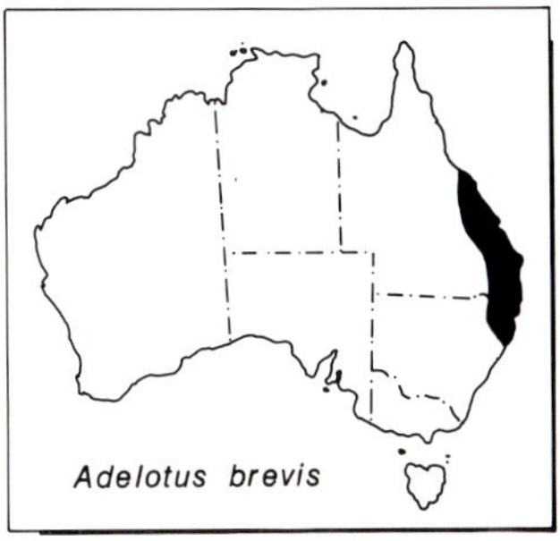

Genus *Rheobatrachus* Liem

The two species in this genus are the only frogs known in which the young develop in the stomach, a process known as gastric brooding. Females stop feeding at reproduction and (presumably) swallow the large fertilized eggs or tadpoles (it has not been observed). Chemical compounds in the eggs and, later, the developing tadpoles, inhibit the production of acid by the stomach wall, which becomes thin-walled. The entire period of development occurs within the stomach "womb", from which juveniles are "born" through the mouth. Following the birth of the last baby, digestive function is restored and the mother recommences feeding.

Rheobatrachus species are entirely aquatic. Their skin is liberally coated with mucus. The fingers are long, slender, pointed and unwebbed and the toes are fully webbed. The eyes are large and projected upwards. Each of the two species is confined to eastern Queensland and each is now considered endangered with *R. silus* probably extinct. Sightings of either species should be reported to the Queensland Department of Environment and Heritage.

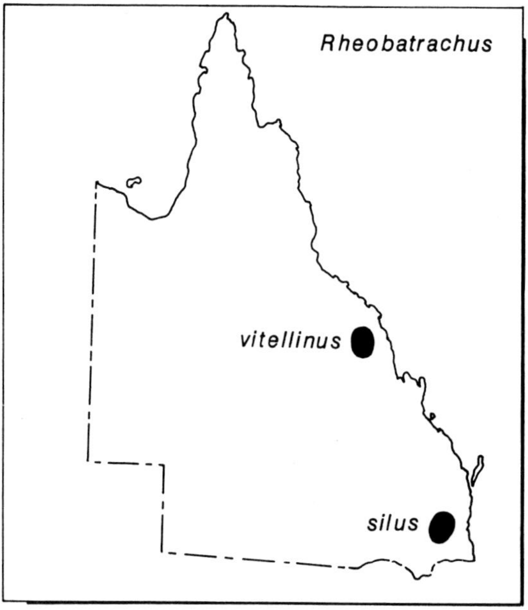

Key to *Rheobatrachus* species

1a. Adult females less than 57 mm SV length; adult males less than 45 mm SV length. Undersurface of body marked with cream or pale yellow .... *R. silus* (p. 351)

b. Adult females greater than 57 mm SV length; adult males greater than 50 mm SV length. Undersurface of body marked with brilliant orange and often slate .... *R. vitellinus* (p. 353)

Rheobatrachus silus

Liem

Southern Gastric Brooding Frog, Southern Platypus Frog

The back ranges from dull grey to slate with obscure darker and lighter patches on it. When the background is pale, a broad brown, and posteriorly curved, supraocular bar is detectable. The undersurface is marked with large patches of cream or pale yellow on white. This species is a strong swimmer but also has the habit of drifting in the water and even of lying on its back.

SIZE: Males 33–41 mm; females 45–54 mm.

HABITAT: Rocky creek beds and adjacent pools in rainforest in the Conondale and Blackall Ranges in southeastern Queensland.

ADVERTISEMENT CALL: A pulsed call with a slight upward inflection and a duration of approximately 0.5 sec, repeated every 6–7 seconds (personal communication from Glen Ingram).

BREEDING AND LARVAE: Eighteen to 25 large cream eggs are swallowed by the female, though neither the process of egg deposition nor their being swallowed has been observed. The unpigmented tadpoles lack tooth rows and do not feed. Oral "birth" takes place after six or seven weeks, the young emerging as fully-formed juveniles.

SIMILAR SPECIES: *Rheobatrachus* is unlikely to be confused with any species except *R. vitellinus* which does not occur in the same area. However, the two species are distinguished readily by differences in coloration and size.

Rheobatrachus silus, Montville, Queensland *(Photo by Gordon Grigg) (Specimen courtesy Glen Ingram).*

Rheobatrachus vitellinus, South Owens Creek, Queensland *(Photo by Mike Mahony).*

Rheobatrachus vitellinus

Mahony, Tyler and Davies

Northern Gastric Brooding Frog,
Northern Platypus Frog

This, the larger of the *Rheobatrachus* species, was reported only in 1984. The eyes are large and protrude conspicuously from the short head. The dorsal skin is irregular and slimy. The fingers are unwebbed and the toes fully webbed. The ventral surface of the body and limbs is marked extensively with a vivid yellow upon a white or slate background. The frogs have been seen basking in the sun in very shallow water.

SIZE: Males 50–53 mm; females 66–79 mm.

HABITAT: Fast flowing rocky creek beds in closed rainforest in the Clarke Ranges near Mackay in central coastal Queensland.

ADVERTISEMENT CALL: Unknown.

BREEDING AND LARVAE: Information is confined to observations of a single female which had 22 young in the stomach. Presumably *R. vitellinus*, like *R. Silus*, broods young in its stomach and exhibits oral "birth".

SIMILAR SPECIES: See *R. silus*.

Genus *Arenophryne* Tyler

The single member of this genus shares with its closest relative, *Myobatrachus gouldii*, the habit of burrowing head-first, using its hands for digging, whereas all other burrowers use their hind feet. *Arenophryne* is confined to coastal sandhills situated between Carnarvon and Geraldton on the Western Australian coast, emerging at night in search of ants which comprise the major component of their diet.

Arenophryne rotunda

Tyler

Sandhill Frog

The body is squat and slightly flattened, with a short and broadly triangular head and extremely short limbs. The fingers are short and broadened to aid the scooping of sand. The skin on the head and body bears large, scattered warts and ridges. The background coloration ranges from grey to pale brown and often has small scarlet and/or bright green patches, and frequently a very narrow, pale mid-dorsal stripe. The lower surfaces are white or dull cream.

SIZE: Males 26–30 mm; females 28–33 mm.

HABITAT: Coarse-grained coastal sandhills. The frogs can be located by the distinctive tracks that they leave on the surface.

ADVERTISEMENT CALL: A short "squelch" emitted above or below the surface.

BREEDING AND LARVAE: Clutches of large, creamy white eggs are laid in March or April at depths of up to 80 cm in moist sand. The larval stages develop entirely within the egg membrane over a period of two months.

SIMILAR SPECIES: *Arenophryne rotunda* resembles slightly the genera *Myobatrachus* and *Pseudophryne*. *Myobatrachus* is also a head-first burrower but has shorter and more muscular limbs, a smooth skin and is substantially larger. *Pseudophryne* species are similar in body form, but do not burrow with their heads, and are usually pigmented boldly in black and white on the undersurfaces.

Arenophryne rotunda, Shark Bay, Western Australia (Photo by Steve Donnellan).

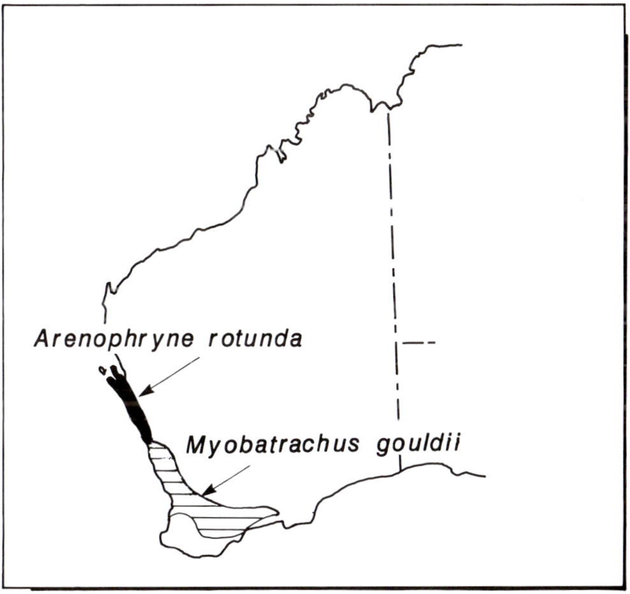

Genus *Myobatrachus* Schlegel

The single species of *Myobatrachus* is one of the most extraordinary frogs in the world. Its overall body form resembles closely that of a mole. The closest relative of this genus is *Arenophryne,* which shares the unusual habit of burrowing head-first rather than excavating a burrow with the feet. *Myobatrachus* is confined to the south-west of Western Australia.

Myobatrachus gouldii

(Gray)

Turtle Frog

The body is gross and bulbous and the head, with its tiny eyes, is distinctly reptilian. The body shape is reminiscent of a pre-spined pouch young echidna! The limbs are short and muscular and there are glands on the inner surface of the upper arm, which may help the male to maintain his grasp on the female by sticking the pair together. The fingers are short and flattened and are involved in digging head-first in soft sand. The colour is slate, pale putty, yellow or rich brown and there are no distinct markings.

SIZE: Males 34–42 mm; females 44–50 mm.

HABITAT: Beneath logs on sandy soils away from water.

ADVERTISEMENT CALL: A series of abrupt, deep croaks.

BREEDING AND LARVAE: Clumps of up to 38 eggs have been found in the soil at depths of more than one metre. The eggs are the largest of any Australian species, up to 7.4 mm in diameter, and development takes place entirely within the egg capsules.

SIMILAR SPECIES: The only other Australian species which can burrow head-first is *Arenophryne rotunda* which is a smaller species with a smooth skin, found further north than the existing known range of *Myobatrachus*.

Myobatrachus gouldii, near Albany, Western Australia *(Photo by Gordon Grigg) (Specimens courtesy Grahame Webb).*

FAMILY MICROHYLIDAE

The family Microhylidae has an extensive geographic distribution, with numerous species in Africa, Asia, South America and New Guinea. In Australia, microhylids are largely confined to the rainforests of northern Queensland and to the high rainfall areas of the extreme north of the Northern Territory.

The Australian microhylid species belong to either *Cophixalus* or *Sphenophryne*. They can be distinguished from one another by features of internal anatomy including the pattern of the superficial muscles of the throat and the arrangement and number of bones and cartilages of the chest, not features that can be used in the field. Rather than produce a key which would be impractical, we have emphasized advertisement calls and geographic distribution as aids to identification. Keys which are based on anatomical features can be found in Zweifel (1985) and Cogger (1992).

The Australian (and New Guinea) microhylid species for which information is available lack free-swimming tadpoles — the tadpole stage being spent inside the egg capsules, away from bodies of water. They all have broad heads and most have the sides of the head darker than the top.

Genus *Sphenophryne* Peters and Doria

The Australian *Sphenophryne* species have squat bodies and short limbs. The tympanum is usually hidden. Finger and toe discs are small or absent and there is no webbing between the digits. The skin is smooth in most species.

Sphenophryne adelphe

Zweifel

Northern Territory Frog

This chunky little frog is morphologically identical with *S. gracilipes* of Cape York. The dorsum is very pale brown or grey with an intense, dark brown stippling on the side of the head becoming sparser posteriorly and ventrally. Most specimens have a faint, narrow pale or dark stripe down the middle of the back. The undersurface is evenly stippled with grey.

SIZE: Males 17–19 mm; females 19–22 mm.

HABITAT: On the Northern Territory mainland this species is restricted to undisturbed areas of paperbark swamp and sandstone outcrops, where it lives under leaf litter. However, on Melville Island north of Darwin the species is abundant around the periphery of pine plantations, and on roadsides and lawns at the settlement of Pickertaramoor.

ADVERTISEMENT CALL: A high pitched "peep-peep-peep-peep . . ." often maintained for more than 10 seconds. Males call mainly from exposed positions on the ground, upon vegetation up to 0.5 metres above the ground, or beneath leaf litter.

BREEDING AND LARVAE: Unknown.

SIMILAR SPECIES: Although physically identical to *S. gracilipes*, there is a broad geographic gap between the distribution of *S. adelphe* in the Northern Territory and *S. gracilipes* of the Cape York Peninsula of Queensland and the adjacent lowlands of southern New Guinea. Within the area of distribution of *S. adelphe*, any confusion with *Uperoleia* species can be resolved by noting the lack of skin glands in *S. adelphe*.

Sphenophryne adelphe, Melville Island, Northern Territory *(Photo by Margaret Davies)*.

Sphenophryne fryi, Cape Tribulation, north Queensland *(Photo by Glenn Hoye)*.

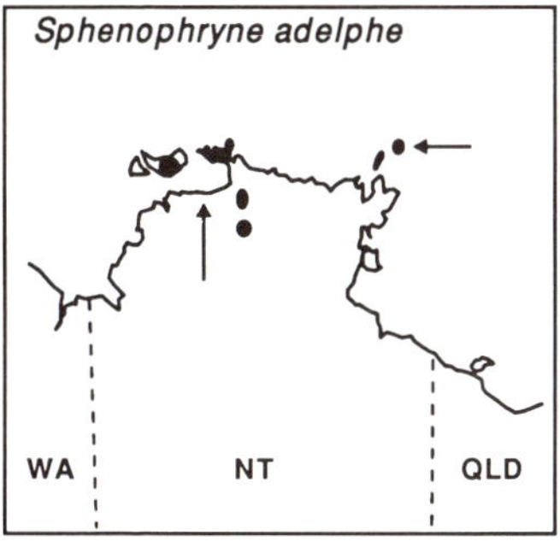

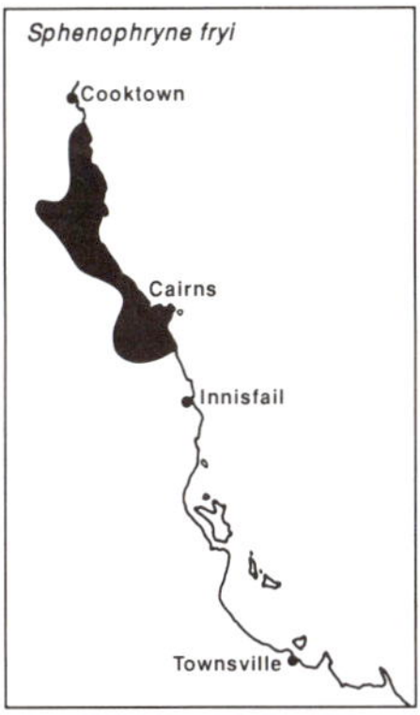

Sphenophryne fryi

Zweifel

Fry's Frog

This species is highly variable in coloration. The dorsum is usually greyish-brown anteriorly merging into a darker reddish-brown posteriorly, with or without dark irregular spots and speckling, and bearing a blackish, or very dark brown, clearly demarcated mask on the side of head. The undersurface is orange coloured and lightly stippled, more dense at the margin of the mandible.

SIZE: Males 23–24 mm; females 23–35 mm.

HABITAT: Beneath debris on the floor of rainforests located between Cooktown and Lambs Range near Atherton in north Queensland.

ADVERTISEMENT CALL: A series of "whistlelike" notes lasting for about two seconds. The call is uttered from beneath debris.

BREEDING AND LARVAE: A clutch of about one dozen eggs is laid beneath debris on damp soil. Single adults have been found with the egg clumps.

SIMILAR SPECIES: *Sphenophryne pluvialis* and *S. robusta* have geographic ranges which overlap that of *S. fryi,* at least in part. Possession of a narrow silvery-white line on the canthus rostralis and the edge of the upper eyelid and reddish eyes distinguishes *S. pluvialis,* whilst the advertisement call of *S. robusta* is a series of short, paired chirps, readily distinguished from the "whistlelike" call of *S. fryi.*

Sphenophryne gracilipes

(Fry)

Slender Frog

The back is a greyish-brown with darker, irregular markings particularly on the sides and flanks. The axilla, groin and sides of the thighs are orange, the undersurface densely stippled on the throat and chest but poorly pigmented on the abdomen.

Sphenophryne gracilipes, Iron Range, north Queensland *(Photo by Owen Kelly)*.

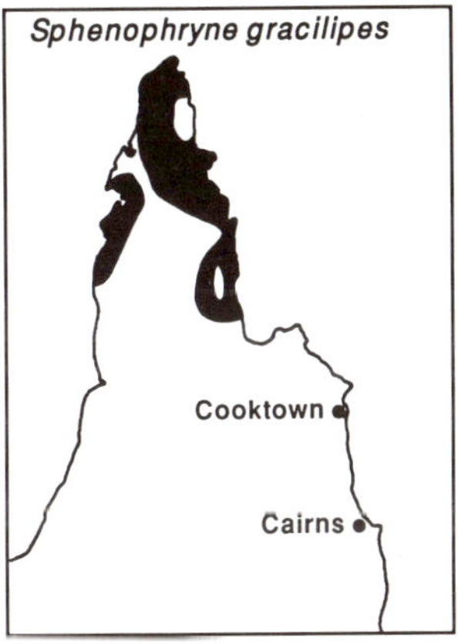

SIZE: Males 17–20 mm; females 19–23 mm.

HABITAT: Grassy woodland, rainforest and lightly forested areas including gallery forests along creeks. The Australian populations extend from just north of Cooktown to Somerset at the northernmost extremity of Cape York Peninsula. It is also found in the lowlands adjacent to the Fly River in New Guinea.

ADVERTISEMENT CALL: Described as a series of high-pitched pips extending over a period of 10–20 seconds. The calling site may be up to 0.5 m above the ground.

BREEDING AND LARVAE: Unknown.

SIMILAR SPECIES: Most similar in size and proportion to *S. adelphe* (see p. 358).

Sphenophryne pluvialis

Zweifel

Rain Frog

The colour of the back of this species ranges from pale greyish-brown with irregular darker suffusion to an intense dark brown with darker markings. In the lightest coloured specimens, the sides of the head and body are considerably darker. There is a white stripe bordering the canthus rostralis and the upper eyelid. The undersurface of the throat and chest are darkly stippled, contrasting with the minimally marked abdomen. The eye is reddish.

SIZE: Males 21–26 mm; females 22–29 mm.

HABITAT: Found in rainforest between Cooktown and Ingham at altitudes ranging from a little above sea-level to 1 300 m.

ADVERTISEMENT CALL: Males call from beneath leaves on the floor of the rainforest. The call is a series of "high-pitched, whistle-like notes", lasting for approximately 1.0–1.5 seconds.

BREEDING AND LARVAE: Unknown.

SIMILAR SPECIES: The geographic distribution overlaps that of *S. robusta* and *S. fryi*. From each of these species *S. pluvialis* is distinguished most readily by the sharp demarcation between the pigmentation of the chest and the abdomen.

Sphenophryne robusta

Fry

Robust Frog

The dorsal colour is dark brown to a dull orange, with diffuse darker markings on the sides of the head. The ventral surface ranges from dull yellow to orange, with diffuse darker markings. A pale mid-vertebral stripe occurs in many individuals.

SIZE: Males 20–27 mm; females 21–33 mm.

HABITAT: Found principally beneath logs and rocks in vine-forest located between Atherton and Paluma, at altitudes above 360 m.

Sphenophryne pluvialis, Streets Creek, Kuranda, north Queensland *(Photo by Gordon Grigg)*.

Sphenophryne robusta, Paluma, north Queensland *(Photo by Glenn Hoye)*.

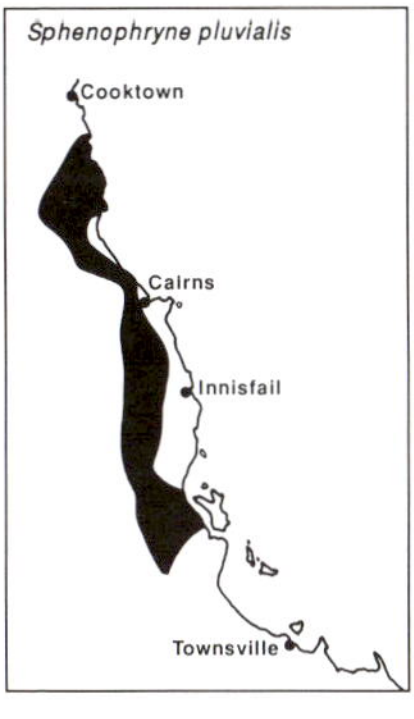

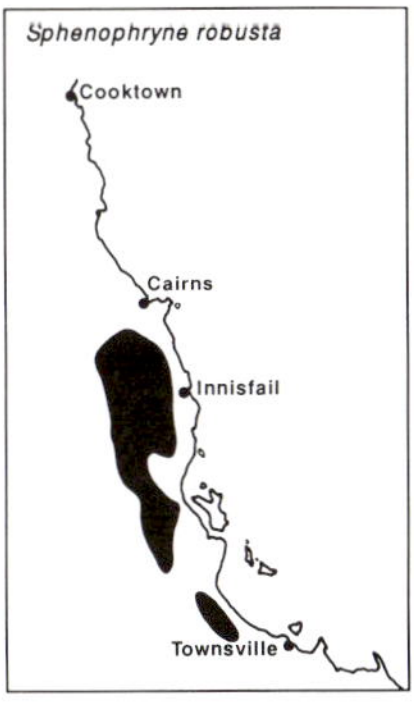

ADVERTISEMENT CALL: A series of high-pitched chirps in repeated couplets emitted from sites beneath cover on the forest floor.

BREEDING AND LARVAE: Unknown.

SIMILAR SPECIES: *Sphenophryne robusta* co-exists with *S. pluvialis* from which the latter can be distinguished by the demarcation of ventral coloration at the junction of the chest and the abdomen.

Genus *Cophixalus* Boettger

Cophixalus species are recognized principally by their loss of clavicle and procoracoid elements of the shoulder girdle, features present in *Sphenophryne*. Many species of *Cophixalus* are known from New Guinea and more await description. Of the 11 species currently recognized in Australia, the majority have been reported in the last 15 years. The current state of knowledge is attributable largely to the pioneering work of R. G. Zweifel, formerly of the American Museum of Natural History in New York.

Cophixalus bombiens

Zweifel

Buzzing Frog

This tiny frog is a dull reddish-brown with intense darker marbling. A faint eyespot in the lumbar region is occasionally present and there may be a bronze mid-vertebral stripe. The undersurface is flecked with grey on a darker background.

SIZE: Males 12–15 mm; females 13–17 mm.

HABITAT: Known only from moist vine-forest on the Windsor Tableland, approximately 50 km south-west of Cape Tribulation.

ADVERTISEMENT CALL: Comparable to the buzz-like sound of an insect, of slightly less than 0.5 seconds duration.

BREEDING AND LARVAE: Unknown.

Cophixalus bombiens, Windsor Tableland, north Queensland *(Photo by Steve Richards).*

Cophixalus concinnus, Carbine Tableland, north Queensland *(Photo by Steve Richards).*

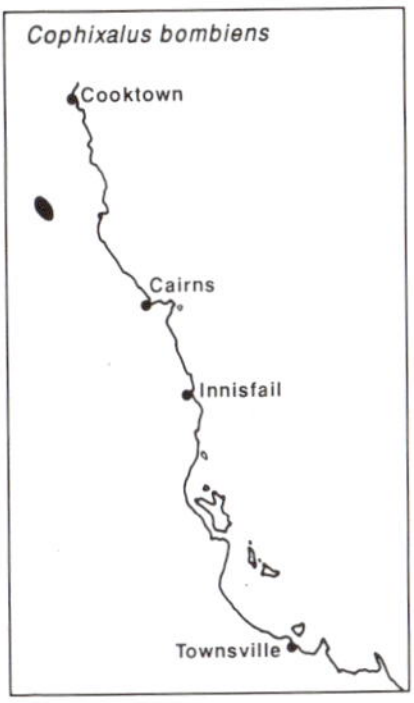

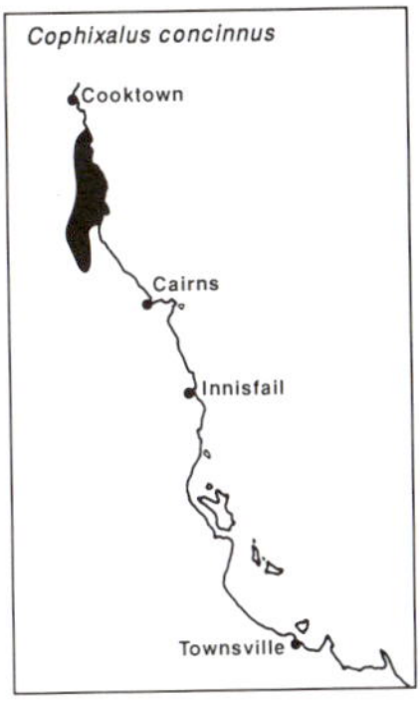

SIMILAR SPECIES: Most similar to *C. exiguus* which is known only from an area between Cape Tribulation and Cooktown. Until more is known about variation in the structure and call of the two species, it would be safer to rely on geographic information to identify these frogs.

Cophixalus concinnus

Tyler

Elegant Frog

The original specimen of *Cophixalus concinnus* is a moderately large frog with large, triangular finger discs and a beautiful and very distinctive coloration. The principal features of this species are moderate to large finger discs and a ventral surface that is commonly yellow or orange, with or without isolated, large, dark brown patches.

SIZE: Males 15–22 mm; females 18–26 mm.

HABITAT: A scansorial species found in rainforests at high elevations between Mossman and Cooktown.

ADVERTISEMENT CALL: A series of clicks or a rattle lasting for about two seconds.

BREEDING AND LARVAE: Unknown.

SIMILAR SPECIES: The only microhylids found with *C. concinnus* are *C. hosmeri* and *C. ornatus*. *Cophixalus hosmeri* is a smaller species with rather longer legs, whilst *C. ornatus* is best distinguished by its unpulsed call.

Cophixalus crepitans

Zweifel

Rattling Frog

This is a particularly small species with moderately long legs and a long rattling call. The discs on the tips of the fingers and toes are moderately developed. The only information on colour in life indicates a burnt orange background colour upon which are scattered brown markings. Undersurface is orange, yellowish-orange or yellowish-green.

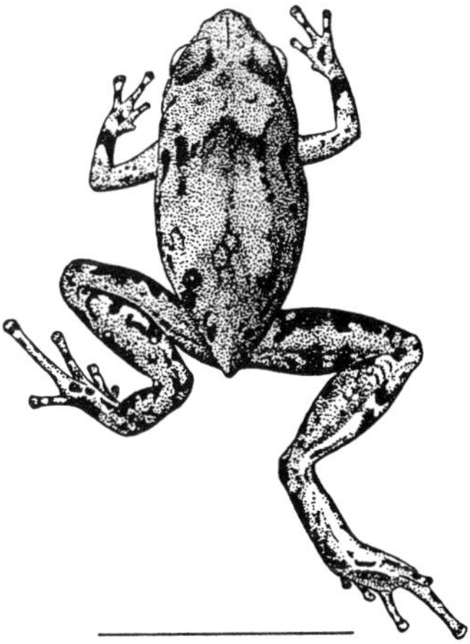

Cophixalus crepitans, QM J28817 (holotype), Leo Creek, McIlwraith Range. Scale line represents 10 mm. *(From Bulletin of American Museum of Natural History, with permission).*

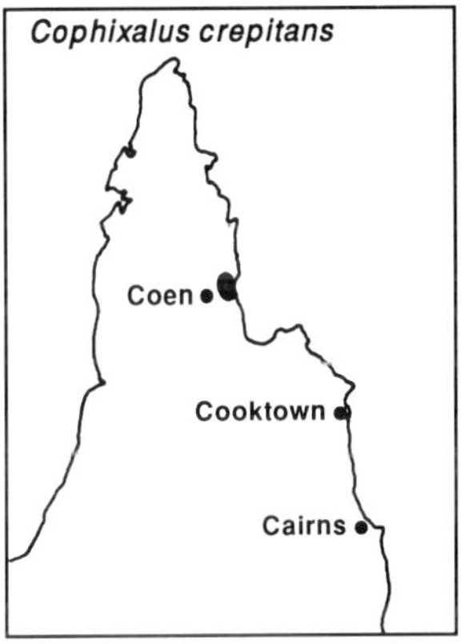

SIZE: Males 12–14 mm; females 14 mm.

HABITAT: Confined to vine forest north-east of Coen.

ADVERTISEMENT CALL: A high-pitched rattling call of about two seconds duration.

BREEDING AND LARVAE: Unknown.

SIMILAR SPECIES: *Cophixalus peninsularis* is the only other species of microhylid believed to exist within the geographic range of *C. crepitans*. The former is a considerably larger species (males 17–18 mm) with proportionately larger discs on the tips of the fingers and toes.

Cophixalus exiguus

Zweifel and Parker

Scanty Frog

This is a small species with a long first finger bearing a large disc. The species is isolated geographically from other species of *Cophixalus*. The background coloration is grey-brown with paler areas including, in some specimens, a broad stripe along the middle of the back and widening between the eyes. The sides of the face may be dark and often there are scattered dark brown patches. There is a deep red patch in the groin.

SIZE: Males 14–16 mm; females 16–19 mm.

HABITAT: Found beneath cover (logs, stones and leaf litter) in rainforest comprising vine forest, including areas with *Acacia*.

ADVERTISEMENT CALL: A tapping call similar to *C. hosmeri* and *C. concinnus* is produced from up to 1.5 m above the ground beneath leaves and within hollow tree trunks.

BREEDING AND LARVAE: Unknown.

SIMILAR SPECIES: *Cophixalus concinnus* is most similar, but appears to have a size range which does not overlap that of *C. exiguus*. After a gap of 25 years it was collected recently on Big Tableland.

Cophixalus hosmeri

Zweifel

Hosmer's Frog

A small, short-legged species which lacks a definite disc on the end of the first finger and has very small discs on the other fingers and toes. The overall coloration can be reddish-brown with obscure darker markings.

SIZE: Males 12–14 mm; females 13–17 mm.

HABITAT: Rainforest at altitudes of 960–1 370 m.

ADVERTISEMENT CALL: Fifteen to 17 clicks emitted over 0.75 to 1.25 seconds. Has been described as a soft tapping call. Males call from vegetation up to one metre above the ground.

Cophixalus exiguus, Big Tableland, north Queensland *(Photo by Harold Cogger)*.

Cophixalus hosmeri, Mt Lewis, north Queensland *(Photo by Mike Mahony)*.

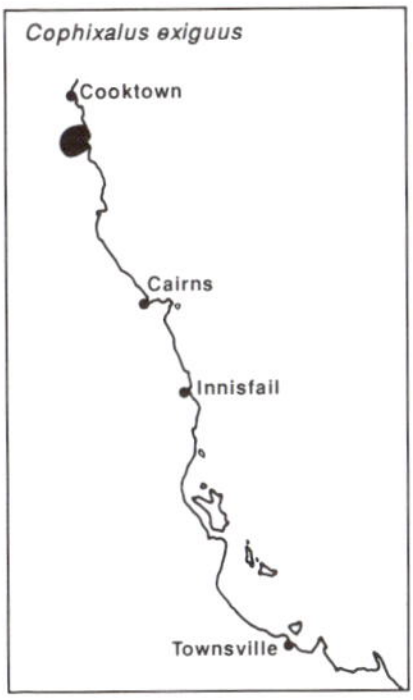

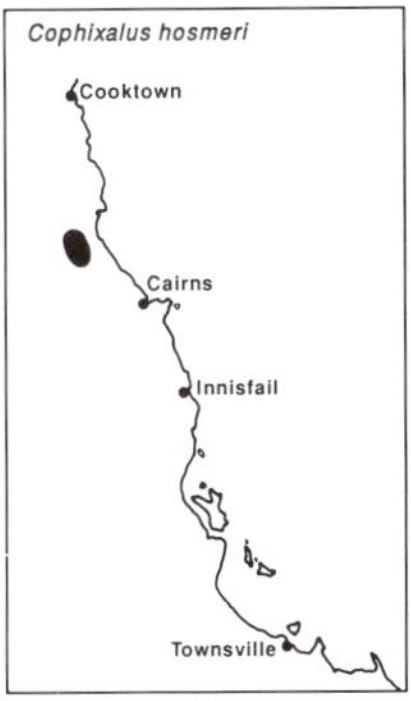

BREEDING AND LARVAE: Unknown.

SIMILAR SPECIES: Confusion is possible with the sympatric species *C. concinnus* and *C. ornatus*. The first is distinctly larger and can be distinguished by size alone, whilst the second is distinguished most readily by its large discs on the fingers and toes.

Cophixalus infacetus

Zweifel

Inelegant Frog

This species has long legs and large discs on all fingers except the first. The head and shoulders are golden brown, changing to dull brown with darker markings posteriorly. A mid-vertebral stripe is present in some individuals.

SIZE: Males 14–16 mm; females 17–18 mm.

HABITAT: Confined to rainforest between Cairns and Ingham.

ADVERTISEMENT CALL: A series of clicks lasting for approximately one second.

BREEDING AND LARVAE: From eight to 13 eggs are laid in a connected string like a bead necklace in a moist situation away from water. The entire period of development takes place within the egg membranes.

SIMILAR SPECIES: The only species co-existing with *C. infacetus* is *C. ornatus* which is a larger species with larger finger and toe discs.

Cophixalus mcdonaldi

Zweifel

McDonald's Frog

A moderately sized species bearing toe discs that are slightly larger than the finger discs. The colour is a dusky golden brown with a dark mark on the side of the head, a narrow brown line behind the eye, a narrow, dark brown fleck on the side and smaller isolated patches on the back and legs. There is a faint yellow patch in the groin.

Cophixalus infacetus, Palmerston, north Queensland *(Photo by Mike Mahony)*.

Cophixalus mcdonaldi, Mt Elliot, north Queensland *(Photo by Steve Richards)*.

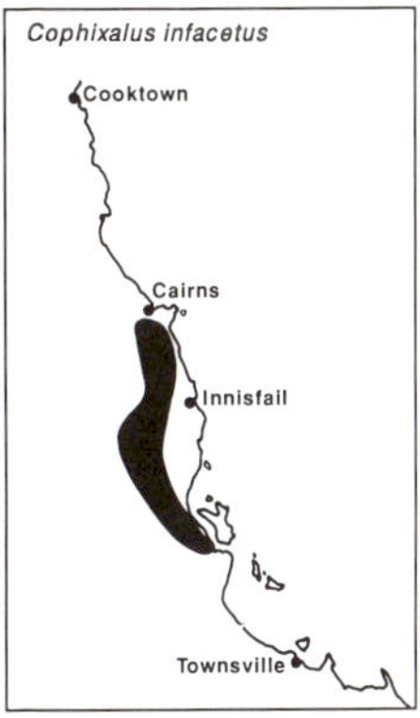

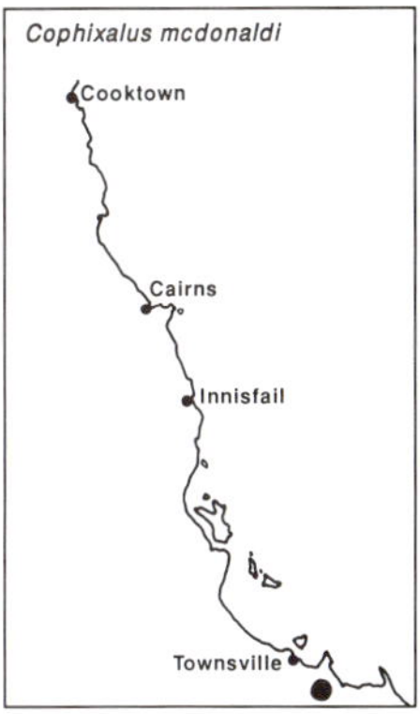

SIZE: Males 16–23 mm; females 21–26 mm.

HABITAT: Seen on palm fronds in rainforest on the slopes of Mt Elliot south-east of Townsville.

ADVERTISEMENT CALL: Unknown.

BREEDING AND LARVAE: Unknown.

SIMILAR SPECIES: No other microhylid overlaps the geographic distribution of this species.

Cophixalus neglectus

Zweifel

Neglected Frog

The distinctive features of *C. neglectus* are its moderate body size, small finger discs and dorsal coloration ranging from brown to orange, and a ventral coloration of yellow or orange.

SIZE: Males 20–33 mm; females 22–29 mm.

HABITAT: Found beneath cover on the floor of high altitude rainforest at approximately 900–1 500 m.

ADVERTISEMENT CALL: A buzzing call with a duration of approximately 0.5 second.

BREEDING AND LARVAE: A group of 14 eggs has been observed attended by an adult frog. The eggs were in three short bead-like chains. The outer capsule is 5.5 mm diameter.

SIMILAR SPECIES: *Cophixalus ornatus* is the only other species of *Cophixalus* known to exist with *C. neglectus*. The former has particularly large finger and toe discs.

Cophixalus ornatus

(Fry)

Ornate Frog

This is a small to moderate-sized species with large and almost triangular finger and toe discs. The colour and pattern of markings are highly variable, with extremes showing mid-vertebral stripes, conspicuous patches, or broad lateral stripes. There tend to be contrasts in colour, with typical components grey, brown, tan and black.

Cophixalus neglectus, Mt Bartle Frere, *(Photo by Andrew Dennis)*.

Cophixalus ornatus, Kuranda, north Queensland *(Photo by Steve Donnellan)*.

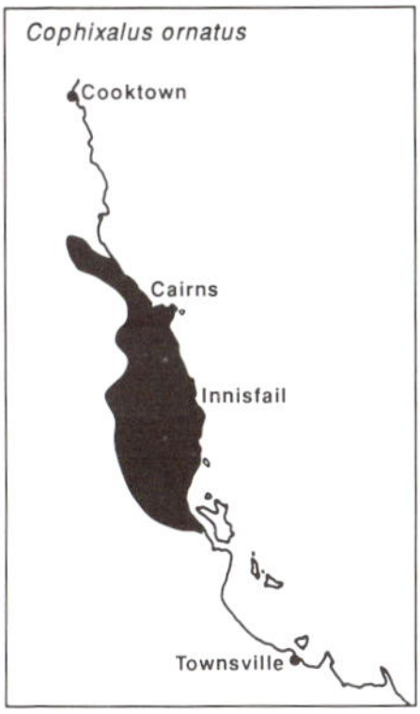

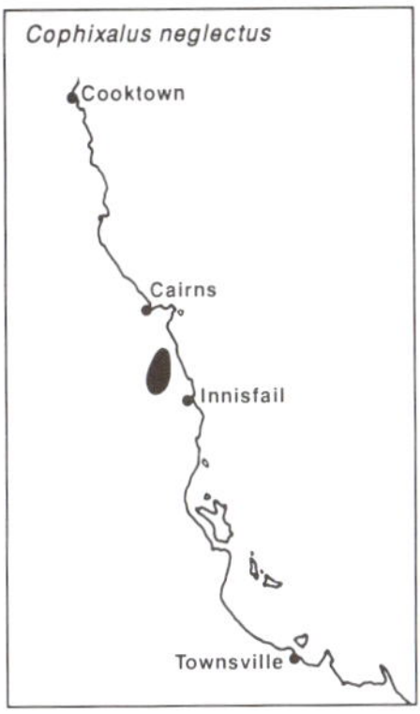

SIZE: Males 17–27 mm; females 22–30 mm.

HABITAT: Found by day beneath ground cover in rainforest between Mossman and Townsville.

ADVERTISEMENT CALL: A short beep of up to 0.5 second duration, emitted from the ground or (more commonly) in vegetation up to 2 m from the ground.

BREEDING AND LARVAE: About 11 eggs are laid in a concealed position. Outer capsule diameter is about 4.3 mm.

SIMILAR SPECIES: The form of the finger discs resembles *C. saxatilis* and *C. concinnus*. The former species is substantially larger, the latter has a clicking call.

Cophixalus peninsularis

Zweifel

Cape York Frog, Peninsula Frog

This species is known only from two specimens taken in a single area of rainforest about 30 km north-east of Coen. It is a moderate-sized species with distinct but not grossly enlarged discs, and a call in the form of a series of clicks lasting for 2.5 seconds.

SIZE: Males 17–18 mm; females unknown.

HABITAT: Found amongst low vegetation at the edge of a creek in rainforest.

ADVERTISEMENT CALL: A series of clicks lasting for 2.5 seconds.

BREEDING AND LARVAE: Unknown.

SIMILAR SPECIES: Within its known geographic range, *C. peninsularis* coexists only with *C. crepitans*. The latter is distinguished readily, being a much smaller species with a maximum length of only 14 mm.

Cophixalus pennisularis, QMJ42061 (holotype); scale line spans 10 mm. *(From Bulletin of American Museum of Natural History with permission.)*

Cophixalus saxatilis, Black Mountain, north Queensland *(Photo by Glenn Hoye).*

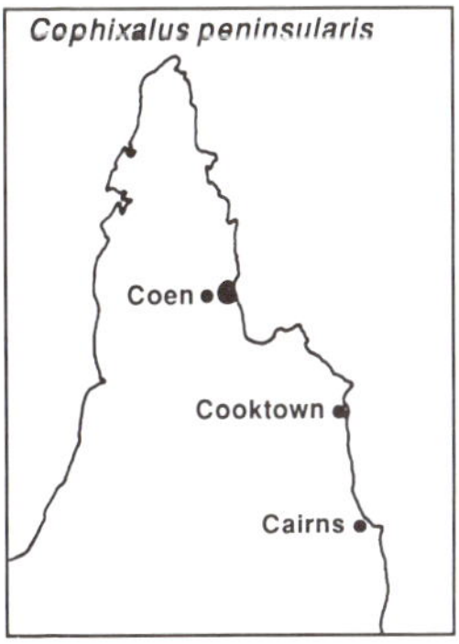

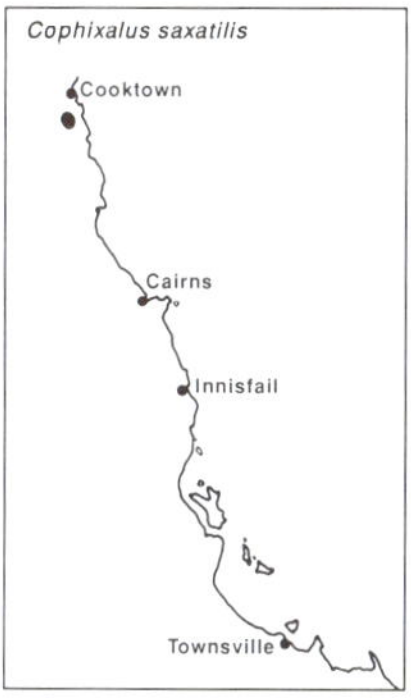

Cophixalus saxatilis

Zweifel and Parker

Rock Frog

This is the largest of the species of *Cophixalus*. It has large triangular finger discs. Females are canary yellow and males are mottled with brown.

SIZE: Males 29–35 mm; females 39–46 mm.

HABITAT: Restricted to an unusual formation of granite boulders on Black Mountain, south of Cooktown, where the frog lives in the caverns formed between enormous boulders.

ADVERTISEMENT CALL: A series of clicks lasting for two to three seconds.

BREEDING AND LARVAE: Unknown.

SIMILAR SPECIES: There are no other species of *Cophixalus* living in the area where *C. saxatilis* can be found, hence there is little possibility of confusion.

FAMILY RANIDAE

The family Ranidae, often called true frogs, is the largest and most widely distributed of all of the anuran families. Australia, with one species, runs second last (ahead of Antarctica) to the other continents of the world in this respect. The only other continental area where the Ranidae is not well represented is the southern half of South America.

As well as being widely dispersed, the Ranidae is also the most diverse family, containing many of the world's most unusual frogs. Included in the family is *Conraua goliath,* the largest of all anurans (body length 30 cm).

The diagnostic characterization of the Ranidae is based upon anatomical structures which are difficult to observe in the living animal. The shoulder girdle is fused at the midline, the digits are usually cylindrical without intercalary structures, and the toes are usually extensively webbed.

Genus *Rana* Linnaeus

For people in many countries *Rana* represents the typical frog: an elongate, long-limbed creature which lives at the edge of the water and leaps for cover at the slightest indication of danger. *Rana* species are dissected in laboratories and consumed in restaurants. In both the Italian and Spanish languages "Rana" is the word for frog.

In contrast with this familiarity, there is only a single species of *Rana* in Australia *(R. daemeli)*, confined to the Cape York Peninsula of north Queensland and the eastern border of Arnhem Land in the Northern Territory. *Rana daemeli* has a more extensive distribution in New Guinea, and its presence in Australia has been attributed to migration across the Torres Strait area during periods when lower sea levels united the land-masses.

Rana daemeli

(Steindachner)

Water Frog, Wood Frog

The head and body are elongate, with the head forming a narrow triangle when viewed from above. The eye is large and the tympanum large and prominent. The limbs are long and muscular and the toes are partially webbed. There is a conspicuous dorso-lateral skin fold from behind the eye to the end of the back, scattered large tubercles on the back and flanks, and a prominent gland at the angle of the jaws. The body and flanks are bronze with a dark stripe from the nostril to the tympanum, and a white stripe along the upper lip. There are dull brown bands upon the arms and legs.

SIZE: Males 43–58 mm; females 58–81 mm.

HABITAT: Amongst dense vegetation at the edge of permanent pools, lakes or streams.

ADVERTISEMENT CALL: A series of seven or eight notes in the form of a low, reedy quacking. *Rana daemeli* is unique amongst Australian species in having paired vocal sacs which project out from the sides of the head like water-wings.

Rana daemeli, Musgrave River, north Queensland *(Photo by Gordon Grigg)*.

Rana daemeli, Musgrave River, north Queensland *(Photo by Gordon Grigg)*.

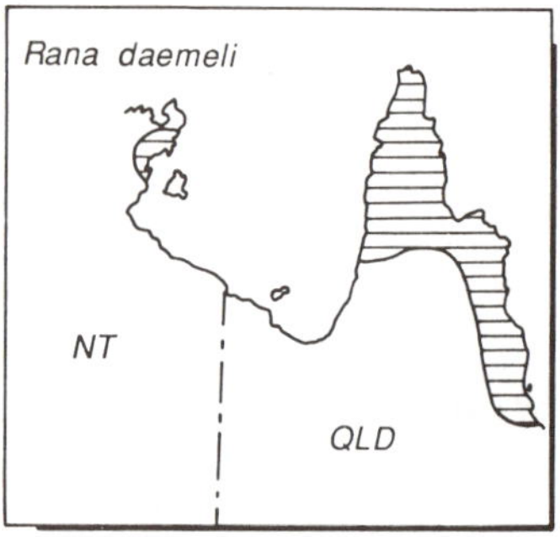

BREEDING AND LARVAE: Several thousand eggs are laid in a large, shapeless mass. The tadpole attains a maximum length of 60 mm. It has a long tail with narrow fins. The back is gold and black and has a faint green suffusion along the midline. The sides are black and the undersurface is grey.

SIMILAR SPECIES: The elongate body form is characteristic of several ground-dwelling *Litoria* species, and a large *L. nasuta* might be mistaken for a small *R. daemeli*. They share elongate skin folds, but in *L. nasuta* the principal fold is not continuous from the eye.

FAMILY BUFONIDAE

Members of the Bufonidae are not native to Australia, but one species, *Bufo marinus,* was introduced in an unsuccessful attempt to control insect pests of sugar cane. It is included in this guide because it has become established. Efforts are currently being made to find a way of controlling toads.

Bufo marinus

(Linnaeus)

Cane Toad

A native of Central and South America, *Bufo marinus* was brought from Hawaii (where it had been introduced) to Australia in 1935 to control two beetle pests of sugar cane. Toads were first released in the Cairns-Innisfail area and, after successes there, further introductions were made in the Ingram, Ayr, Mackay and Bundaberg areas. After almost 60 years, *Bufo marinus* has spread south, north and west. The southern limit at present is near Coffs Harbour whilst in the north it extends 100 km north of Borroloola in the Northern Territory. This remarkable dispersal is due to the toads' great adaptability, for they are able to tolerate and even thrive in virtually any habitat. Another contributing factor to this "population explosion" is the toxic properties of the secretions from the large

Bufo marinus, Dorunda, north Queensland *(Photo by Gordon Grigg).*

parotoid glands which make toads a most unpalatable meal. Only a few species of birds are known to feed on them. Snakes, the greatest predator of amphibians, are sometimes found dead with a toad part way down the throat.

The dorsal surface is pale brown with dark brown blotches. Many large warts cover the back, particularly just behind the head where the large swollen parotoid glands are found. The ventral surface is cream and granular and may be blotched in shades from tan to black. The head is large, short and rounded with protruding eyes and thick warty eyelids. The pupil is horizontal and the iris golden with a network of thin black lines throughout. No teeth are present. The short and stubby fingers are devoid of web while the toes have a fleshy web at their base.

SIZE: Males 75–110 mm; females 90–250 mm.

HABITAT: *Bufo marinus* has been found in a wide variety of habitats, preferring tracks, roads and low grassland. Dense vegetation tends to create a barrier to their dispersal.

ADVERTISEMENT CALL: A rapid "pop-pop-pop" sounding like a small engine or generator.

BREEDING AND LARVAE: Masses of eggs in long chains can be deposited in almost any body of water during spring and summer. A female is capable of laying 35 000 eggs in one season. The small black tadpoles take about 12 weeks to metamorphose, after which vast numbers of small toads, about 10 mm long, may be seen dispersing from the ponds.

SIMILAR SPECIES: The adult is sufficiently distinct from native species to minimize confusion. However, the large *Cyclorana* species *C. australis* and *C. novaehollandiae* have been mistaken by some naturalists. The vast parotoid glands of *B. marinus* are diagnostic. Confusion of juvenile toads with small native species such as *Uperoleia* species is possible, but the absence of the colourful patches in the groin and behind the thighs will separate them.

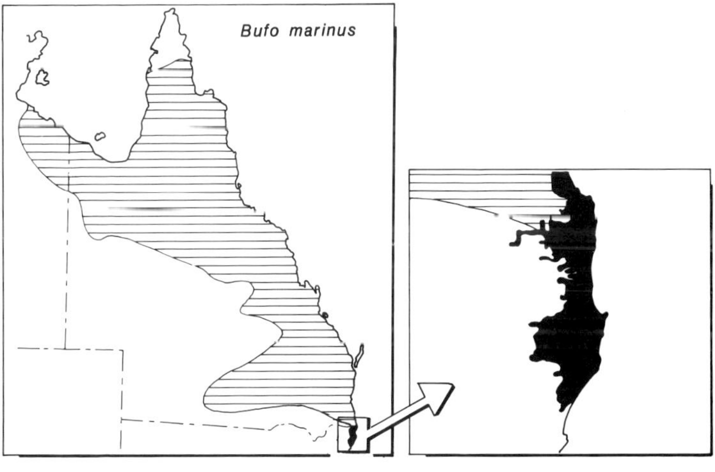

Chapter 4

Collecting, Keeping and Photographing Frogs

For the enthusiastic naturalist it will not be sufficient to just read about animals, he or she will want to observe and study them in their natural environment.

The naturalist studying frogs for the first time may be unsure of how to go about it. We have spent many years and travelled thousands and thousands of kilometres tracking down these little beasts and some of the experience we have gained may be useful to the beginner.

From the preceding chapters it is seen that frogs are found in nearly any habitat, but they are most easily found near water. It is worth remembering that nearly all species are nocturnal and that most are active during the warmer spring and summer months, especially after rain. From these points, the pessimist may deduce that if frogs are to be collected it will have to be on pitch black nights in mosquito and snake-infested swamps in the pouring rain!

Equipment

Only the bare essentials are necessary — a torch, gum boots and something for putting captives in. A powerful torch is essential and it should be rugged and waterproof. Head torches are also very good because they leave both hands free. Wide-mouthed jars or plastic containers with snap-on lids are suitable for field-storage of the specimens, but thin polyethylene bags (about 20 cm by 30 cm) are probably better than anything else. Before use, these are carried easily in the pockets. After placing the frog in the bag, blow air into the bag and seal the neck with a rubber band or by tying the neck. Polyethylene bags prevent frogs from drying out and we find that under these conditions frogs live well as long as they are kept in a cool place.

Legislative restrictions

Frogs are "protected animals" to some extent in all States and Territories except South Australia. The legislation is modified from time to time, and it is essential that anyone wishing to collect or keep frogs checks the current requirements from the appropriate State or Territory authority — the National Parks and Wildlife Service or Department of Environment. There are also controls on the transport of frogs between States or Territories, whilst the transport of live or preserved frogs into or out of Australia is permitted only under exceptional circumstances.

The Cane Toad is a declared "prohibited species" in most States and permission to import for scientific or teaching purposes is granted only to recognized institutions, under permit.

Collecting hints

Commonly, the best indication that frogs are present in an area may be the advertisement calls of the males. It is difficult to determine the position of a calling frog, and a technique known as triangulation may be used. This requires three people (two may be sufficient) standing several metres apart. Each shines a torch on the spot where they think the call originates. The approximate location of the frog is indicated where the beams cross. This method is particularly suitable for small terrestrial frogs (e.g., *Crinia*) which hide in grass clumps or leaf litter.

When collecting during the day, the most successful method is to walk down a shallow creek bed and turn over stones and logs (being sure to replace them exactly as they were). Most frogs rely on their camouflage for survival and will not move to escape until the last moment. Creep as close as possible and pick up the frog quickly and securely. If larger frogs (e.g., *Mixophyes* species) are held around the waist with the legs extended they cannot escape.

It is often very easy to collect by driving at night along a quiet road during or just after rain. Select one running along a creek or swamp or crossing a number of streams. If the road passes through different habitats, such as rainforest, open forest, and cleared country, a large diversity of frogs may be collected. While driving slowly (10–15 km/h) frogs will be easily seen sitting on the road. On a road near Ourimbah,

New South Wales, John Barker and Gordon Grigg once collected 11 species in about an hour and a half, and that was in May at the end of the season.

It is essential to keep accurate and detailed field notes recording the exact locality where a frog was collected, and also the time, date, weather conditions, temperature, and other relevant information. One of our data sheets is reproduced on page 386. We find a small hand-held tape recorder ideal for recording notes which we transcribe into a field notebook the following morning.

If the frog is calling when collected, note the position of the animal in relation to the water and describe the call as best you can. With the advent of small hand-held cassette recorders, a tape recording of the call can be made easily, and compared with commercially recorded calls. Recordings suitable for scientific analysis can be made only on the finest quality equipment with the microphone held very close (5–15 cm) to the frog. Interesting and valuable information can be gathered by any naturalist, and observations or specimens which seem unusual should be forwarded to museums or other institutions. It is still possible to find new species, especially in the more remote areas, and there are many gaps in our knowledge of life cycles and habits.

Keeping frogs in captivity

Adult frogs are not difficult to keep in captivity and can easily be housed in an aquarium. Try to lay out the aquarium with a habitat suiting the particular frogs, for example, about 10 cm of sand for burrowing frogs, and branches or other vegetation for tree frogs. Always include some stones for cover by day and provide water in a dish or trough sunk into the sand in a corner. Fly-screen mesh or cotton close-weave material tacked to a wooden frame are the best coverings for the tank. Frogs will eat only live food, so a good supply of mealworms, earthworms, flies, beetles and ants will have to be made available. Some of the small *Pseudophryne* species are unable to catch flies or other fast-moving prey and their natural diet

DATA SHEET (AMPHIBIA)

Use new sheet for each species at each locality.

Trip: SYDNEY - E. GIPPSLAND - MT BAW BAW - SYDNEY

Date: 7-11-72 Time: 1500

Species: L. RANIFORMIS Number collected: 12

Bag No. 2 (amplexing pair), 3, 4, 5, 6, 7, 9-15.

Where seen: amplexing pair in water, rest caught under logs.

Locality: 6 M Sth BOMBALA on way to Delegate

Habitat: Farm Dam c̄ Reeds

Abundance: V. Common

Calling? Yes Other calls? none.

Notes on call: recorded .... daytime calling!

Recording notes: Tape 12 Side A

Weather: Wind? light Rain? none % Cloud 0

Temperatures: Frog — Wet bulb 13 Dry bulb 20-19

Water 24°C Relative humidity: 46%

Rainfall: Past 24 hours none Past 7 days 3" last 3 days

Photographed? Several Catalogue No.

Comments: There were many raniformis active in pond in bright sunshine. Many amplexing pairs. Vigorous activity. When 2 plovers flew over they all dived.

consists chiefly of ants, termites and small beetles. Remember to separate frogs according to size; we have lost small frogs to the larger, voracious *Litoria aurea* and *Cyclorana australis*. Furthermore, the aquarium should not be overcrowded — three or four medium-sized frogs in a 40 cm × 20 cm × 20 cm tank is enough.

Collecting and keeping tadpoles

Knowledge of the larval life histories of frogs is remarkably scant. Angus Martin, of the University of Melbourne, has compiled an excellent key to the tadpoles of the Melbourne area, and Michael Tyler, Graeme Cook and Margaret Davies have prepared one on the species of the Northern Territory floodplain, but information on other species is either lacking or not easily available. A list of all tadpoles that have been described is in Tyler (1994). The oral disc of tadpoles is the main taxonomic character, because most species have a particular arrangement of labial teeth and papillae. A 10× lens is needed for examination of the mouthparts.

Tadpoles may be collected with a dip net and should be kept in the pond water from which they came. If using tap water, let it stand in a bucket for 24 hours before use, to eliminate chlorine. If studies on comparative development are to be done, tadpoles will have to be examined and kept in separate tanks according to species. Tadpoles can be fed boiled lettuce and, if the tank can be kept where it gets sunlight for part of the day, a good growth of algae will obviate the need for other feeding. When the tadpoles have nearly completed metamorphosis they should be transferred to a container with very shallow water and a stone or two for the froglets to sit on when they leave the water. As with adult frogs, overcrowding must be avoided.

Preserving frogs and tadpoles

If specimens are to be preserved for reference or for sending to museums, 10 per cent formalin or 65 per cent alcohol should be used. For tadpoles, 5–7 per cent formalin or 65 per cent alcohol is suitable, although the ideal preservative is Tyler's fixative (10 ml concentrated formalin, 4 ml of 10 per cent aqueous solution of calcium chloride, 0.2 g cobalt nitrate, 90 ml water). Most of the colours will fade or change after preservation, so notes on live colour must be kept or a photograph made. Specimens should be stored in screw-cap jars with labels attached to each frog, recording species, date, and locality. We favour the use of an ink pen such as the fine Mitsubishi Uniball because it is waterproof.

Photography

Because the colours of frogs, often so striking in life, usually fade when a specimen is preserved, a good colour photograph can accurately record the living appearance. Since the identification of frogs often depends on coloration in life, a photographic record can be very important. However, there need be no further justification for photographing frogs beyond an aesthetic one, for their form, texture, and coloration place them among the most photogenic of our fauna.

The aim of a natural history photographer is usually to produce an attractive and characteristic photograph depicting the animal in natural surroundings. The best photos are undoubtedly those taken in the field. We have endeavoured to do this wherever possible and many of the frogs photographed for this book have not been removed from the spot where they were sighted. If this is not possible, the photographer can collect the frog and also some bits and pieces from the habitat to mock up a "natural" setting at home, or can photograph the frog on some plain background. The latter course is less pleasing from an artistic viewpoint but is certainly adequate for scientific purposes. Detail of fingers and toes may be specially clear against a plain background. Frogs often look best with pupils dilated, as at night. Many frogs are active and reluctant to "pose", and patience is perhaps the main requirement for any natural history photographer.

Apart from considerations of background and setting, two difficulties confront the frog photographer. These are the small size of many frogs, and the need to provide enough illumination from the right directions. A single lens reflex camera fitted with extension bellows, extension tubes, macro lens, or supplementary close-up lens allows the image to almost fill the viewfinder, thus coping with the first problem. Such a camera-lens combination has, however, a very shallow depth of field and so one must provide enough illumination to make exposures using a very small aperture such as f/16 or f/22. Natural lighting is inadequate for this in most cases and so a flash unit is usually employed. The position of the flash unit is very important for it must be at an angle sufficient to provide shadowing and a three-dimensional image without casting large black pools of shade in which no detail is visible. For this reason a fill-in flash is required for top quality pictures. A system that we have found ideal in the field and in the laboratory consists of two low-power electronic flash units, one mounted on each side of the camera with one of them (the fill-in) "muzzled" with a 0.2 neutral density filter. The flashes are attached to a cross-bar by ball-and-socket mounts. Discharge of both units is synchronized with the opening of the shutter. Using this system with a macro lens and ASA 64 slide film, we find f/16 or f/22 is satisfactory and provides adequate depth of field. The double-flash combination provides a picture in which the subject is well "moulded" by the interplay of light and shade.

A good place for frogs. Near Glen Lyon, south-east Queensland.

Selected Bibliography

Identification and biology

Barker, J. and Grigg, G. C., 1977. A Field Guide to Australian Frogs. Rigby: Adelaide. 229 pp.

Brook, A. J., 1983. Atlas of Australian Anura. Dept Zoology, Univ. Melbourne Publ. 7.

Cogger, H. G., 1992. Reptiles and amphibians of Australia. 5th Edition, Reed: Sydney.

Cogger, H. G., Cameron, E. E. and Cogger, H. M., 1983. Zoological Catalogue of Australia, Vol. 1, Amphibia and Reptilia. Australian Government Printing Service: Canberra. 313 pp.

Duellman, W. E. and Trueb, L., 1986. Biology of Amphibians. McGraw-Hill: New York. 670 pp.

Friend, G. R. and Cellier, K. M., 1990. Wetland herpetofauna of Kakadu National Park, Australia: seasonal richness trends and the effects of feral ungulates. *J. Trop. Ecol.* **6:** 131–52.

Frost, D. R. (ed.), 1985. Amphibian Species of the World; a taxonomic and geographical reference. Allen Press and Assoc. Systematics Collections: Lawrence, Kansas. v + 732 pp.

Glasby, C. J., Ross, G. J. B. and Beasley, P. L. (eds), 1993. Fauna of Australia, Vol. 2A Amphibia and Reptilia. Australian Government Printing Service. 439 pp.

Grigg, G. C. and Barker, J., 1973. Explanatory Notes to Frog Calls of South Eastern Australia. G. Grigg and J. Barker: Sydney. 9 pp (with audio cassette of frog calls).

Grigg, G. C., Shine R. and Ehmann, H., 1985. The Biology of Australasian Frogs and Reptiles. Surrey Beatty & Sons: Chipping Norton. 527 pp.

Hero, J.-M., Littlejohn, M. and Marantelli, G., 1991. Frogwatch Field Guide to Victorian Frogs. Department of Conservation and Environment: Victoria.

Kikkawa, J., Monteith, G. B. and Ingram, G. J., 1981. Cape York Peninsula; major region of faunal interchange. Pp. 1695–742 *in* Ecological Biogeography of Australia ed by A. Keast. Dr W. Junk: The Hague.

Littlejohn, M. J. and Martin, A. A., 1974. The amphibia of Tasmania. *In* Biogeography and Ecology in Tasmania ed by W. D. Williams. W. Junk: The Hague.

McDonald, K. R., 1992. Distribution patterns and conservation status of north Queensland rainforest frogs. *Conservation Technical Report* No.1, Queensland Department of Environment and Heritage.

Main, A. R., 1965. Frogs of southern Western Australia. *West Aust. Nat. Club Handbook* **8:** 1–73.

Martin, A. A. and Littlejohn, M. J., 1982. Tasmanian amphibians. Fauna of Tasmania Handbook No. 6. Fauna of Tasmanian Committee, University of Tasmania: Hobart.

Moore, J. A., 1961. The frogs of eastern New South Wales. *Bull. Amer. Mus. Nat. Hist.* **121**(3): 302.

Richards, S. J. and James, C., 1992. Foot-flagging displays of some Australian frogs. *Mem. Qld Mus.* **32**(1): 302.

Robinson, M., 1993. A Field Guide to Frogs. Australian Museum and Reed: Sydney.

Savage, J. M., 1973. The geographic distribution of frogs: patterns and predictions. Pp. 351–445 *in* Evolutionary Biology of the Anurans. University of Missouri Press: Columbia, Missouri.

Tyler, M. J., 1978. Amphibians of South Australia. South Australian Government Printer: Adelaide. 84 pp.

Tyler, M. J., 1982. Frogs. Collins: Sydney. (Australian Naturalists Library). 256 pp. (2nd ed.).

Tyler, M. J., 1985. Reproductive modes in Australian amphibia. Pp. 265–67 *in* The Biology of Australasian Frogs and Reptiles ed by G. C. Grigg, R. Shine and H. F. W. Ehmann. Surrey Beatty & Sons: Chipping Norton with Royal Zoological Society of NSW, Mosman.

Tyler, M. J., 1992. Encyclopedia of Australian Animals. Frogs. Collins, Angus and Robertson: Pymble, New South Wales.

Tyler, M. J., 1994. Australian Frogs: A Natural History. Reed Publications: Sydney.

Tyler, M. J., Crook, A. G. and Davies, M., 1983. Reproductive biology of the frogs of the Magela Creek System, Northern Territory, Australia. *Rec. S. Aust. Mus.* **18**(18): 415–40.

Tyler, M. J. and Davies, M., 1986. Frogs of the Northern Territory. Conservation Commission of the Northern Territory: Alice Springs. vi + 7 pp.

Tyler, M. J., Davies, M. and Watson, G. F., 1991. The frog fauna of Melville Island, Northern Territory. *The Beagle* **8**(1): 1–10.

Tyler, M. J., Smith, L. A. and Johnstone, R. E., 1994. Frogs of Western Australia (Revised Edition). Western Australian Museum: Perth.

Watson, G. F. and Littlejohn, M. J., 1985. Patterns of distribution, speciation and vicariance biogeography of south eastern Australian amphibians. Pp. 91–97 *in* The Biology of Australasian Frogs and Reptiles ed by G. C. Grigg, R. Shine and H. F. W. Ehmann. Surrey Beatty & Sons: Chipping Norton with Royal Zoological Society of NSW: Mosman.

Family Myobatrachidae

Blake, A. J. D., 1973. Taxonomy and relationships of myobatrachine frogs (Leptodactylidae): a numerical approach. *Aust. J. Zool.* **21**: 119–49.

Corben, C. and Ingram, G. J., 1987. A new barred river frog (Myobatrachidae: *Mixophyes*). *Mem. Qld Mus.* **26**(1): 233–37.

Czechura, G. V., 1986. A new species of *Taudactylus* (Myobatrachidae) from south eastern Queensland, Australia. *Mem. Qld Mus.* **22**(2): 299–307.

Czechura, G. V., 1986. Kroombit Tops torrent frog *Taudactylus pleione*, with a key to the species of *Taudactylus*. *Qld Nat.* **27**(1–4): 68–71.

Czechura, G. V. and Ingram, G. J., 1990. *Taudactylus diurnus* and the case of the disappearing frogs. *Mem. Qld Mus.* **29**(2): 361–65.

Davies, M., 1991. Descriptions of the tadpoles of some Australian limnodynastine leptodactylid frogs. *Trans. Roy. Soc. S. Aust.* **115**(2): 67–76.

Davies, M., 1992. Early development of *Limnodynastes terraereginae* and *L. fletcheri* (Anura: Leptodactylidae: Limnodynastinae). *Trans. Roy. Soc. S. Aust.* **116**(4): 117–22.

Davies, M. and Littlejohn, M. J., 1986. Frogs of the genus *Uperoleia* Gray (Anura: Leptodactylidae) in south-eastern Australia. *Trans. Roy. Soc. S. Aust.* **109**(3): 111–43.

Davies, M. and McDonald, K. R., 1985. A redefinition of *Uperoleia rugosa* (Andersson) (Anura: Leptodactylidae). *Trans. Roy. Soc. S. Aust.* **109**(2): 37–42.

Davies, M., Mahony, M. and Roberts, J. D., 1985. A new species of *Uperoleia* (Anura: Leptodactylidae) from the Pilbara region, Western Australia. *Trans. Roy. Soc. S. Aust.* **109**(3–4): 103–08.

Davies, M., McDonald, K. R. and Corben, C., 1986. The genus *Uperoleia* Gray (Anura: Leptodactylidae) in Queensland, Australia. *Proc. Roy. Soc. Vic.* **98**(4): 147–88.

Davies, M., Watson, G. F. and Miller, C. A., 1987. New records of *Uperoleia* (Anura: Leptodactylidae) from Western Australia with supplementary osteological data on *Uperoleia micromeles*. *Trans. Roy. Soc. S. Aust.* **111**(4): 202–02.

de Bavay, J. M., 1993. The developmental stages of the Sphagnum Frog, *Kyrranus sphagnicolus* Moore (Anura: Myobatrachidae). *Aust. J. Zool.* **41**: 151–201.

Ehmann, H. F. W. and Swan, G., 1985. Reproduction and development in the marsupial frog, *Assa darlingtoni* (Leptodactylidae: Anura). Pp. 279–85 *in* The Biology of Australasian Frogs and Reptiles ed by G. C. Grigg, R. Shine and H. F. Ehmann. Surrey Beatty & Sons: Chipping Norton with Royal Zoological Society of NSW: Mosman.

Gollman, B., 1991. A developmental table of *Crinia signifera* Girard, 1853 (Anura: Myobatrachinae) *Alytes* **9**(2): 51–58.

Gollmann, B. and Gollmann, G., 1991. Embryonic development of the myobatrachine frogs *Geocrinia laevis*, *Geocrinia victoriana*, and their natural hybrids. *Amphibia — Reptilia* **12**: 103–10.

Heyer, W. R. and Liem, D. S., 1976. Analysis of the intergeneric relationships of the Australian frog family Myobatrachidae. *Smithsonian Cont. Zool.* **233**: 1–29.

Hutchinson, M. N. and Maxson, L. R., 1987. Albumin evolution, paraphyly and the relationships of the Australian frog *Megistolotis lignarius* (Anura: Limnodynastinae). *Syst. Zool.* **36**(4): 387–94.

Ingram, G., 1980. A new frog of the genus *Taudactylus* (Myobatrachidae) from mid-eastern Queensland with notes on the other species of the genus. *Mem. Qld Mus.* **20**(1): 111–19.

Ingram, G. and McDonald, K. R., 1993. An update on the decline of Queensland frogs. *In* Herpetology in Australia ed by D. Lunney and D. Ayers. *Trans. Roy. Zool. Soc. of NSW.*

Lee, A. K., 1967. Studies in Australian Amphibia. II. Taxonomy, ecology, and evolution of the genus *Heleioporus* Gray (Anura, Leptodactylidae). *Aust. J. Zool.* **15:** 367–39.

Leong, A. S.-Y., Tyler, M. J. and Shearman, D. J. C., 1986. Gastric brooding: a new form in a recently discovered Australian frog of the genus *Rheobatrachus. Aust. J. Zool.* **324:** 205–09.

Littlejohn, M. J. and Martin, A. A., 1965. The vertebrate fauna of the Bass Strait Islands: 1. The Amphibia of Flinders and King Islands. *Proc. Roy. Soc. Vic.* **79**(1): 247–56.

Littlejohn, M. J., Harrison, P. A. and MacNally, R. C., 1985. Interspecific acoustic interactions in sympatric populations of *Ranidella signifera* and *R. parinsignifera* (Anura: Leptodactylidae). Pp. 287–96 *in* The Biology of Australasian Frogs and Reptiles ed by G. C. Grigg, R. Shine and H. F. W. Ehmann. Surrey Beatty and Sons: Chipping Norton with Royal Zoological Society of NSW; Mosman.

Lynch, J. D., 1971. Evolutionary relationships, osteology, and zoogeography of leptodactyloid frogs. *Misc. Publ. Univ. Kansas Mus. Nat. Hist.* **53:** 1–238.

MacNally, R. C., 1985. Habitat and microhabitat distributions in relation to ecological overlap in two species of *Ranidella* (Anura). *Aust. J. Zool.* **33:** 329–38.

Mahony, M. J. and Roberts, J. D., 1986. Two new species of desert burrowing frogs of the genus *Neobatrachus* (Anura: Myobatrachidae) from Western Australia. *Rec. West. Aust. Mus.* **13**(1): 155–70.

Mahony, M. J., Tyler, M. J. and Davies, M., 1984. A new species of the genus *Rheobatrachus* (Anura: Leptodactylidae) from Queensland. *Trans. Roy. Soc. S. Aust.* **108**(3): 155–62.

Main, A. R., 1957. Studies in Australian Amphibia. 1. The genus *Crinia* Tschudi in south western Australia and some species from south eastern Australia. *Aust. J. Zool.* **5:** 30–55.

Malone, B. S., 1985. Mortality during the early life history stages of the Baw Baw frog, *Philoria frosti* (Anura: Myobatrachidae). Pp. 1–5 *in* The Biology of Australasian Frogs and Reptiles ed by G. C. Grigg, R. Shine and H. F. W. Ehmann. Surrey Beatty & Sons: Chipping Norton with Royal Zoological Society of NSW: Mosman.

Martin, A. A., 1972. Studies in Australian Amphibia. III. The *Limnodynastes dorsalis* complex (Anura: Leptodactylidae). *Aust. J. Zool.* **20:** 165–211.

Odendaal, F. J., Bull, C. M. and Telford, S. R., 1983. The vocabulary of calls of *Ranidella riparia* (Anura: Leptodactylidae). *Copeia* **1983**(2)**:** 534–37.

Osborne, W. S., 1989. Distribution, relative abundance and conservation status of corroboree frogs, *Pseudophryne corroboree* Moore (Anura: Myobatrachidae). *Aust. Wildl. Res.* **16:** 537–47.

Parker, H. W., 1940. The Australasian frogs of the family Leptodactylidae. *Novit. Zool.* **42:** 1–106.

Roberts, J. D., 1984. Terrestrial egg deposition and direct development in *Arenophryne rotunda* Tyler, a myobatrachid frog from coastal sand dunes at Shark Bay, WA. *Aust. Wildl. Res.* **11**(1)**:** 191–200.

Roberts, J. D., 1985. Population density estimates for *Arenophryne rotunda*: is the Round Frog rare? Pp. 463–67 *in* The Biology of Australasian Frogs and Reptiles ed by G. C. Grigg, R. Shine and H. F. W. Ehmann. Surrey Beatty & Sons: Chipping Norton with Royal Zoological Society of NSW: Mosman.

Roberts, J., Mahony, M., Kendrick, P. and Majors, C. M., 1991. A new species of burrowing frog, *Neobatrachus* (Anura: Myobatrachidae), from the eastern wheatbelt of Western Australia. *Rec. W. Aust. Mus.* **15** (1)**:** 23–32.

Roberts, J. D. and Maxson, L. R., 1986. Phylogenetic relationships in the genus *Limnodynastes* (Anura: Myobatrachidae): a molecular perspective. *Aust. J. Zool.* **34:** 561–73.

Roberts, J. D. and Seymour, R. S., 1989. Non-foamy egg masses in *Limnodynastes tasmaniensis* (Anura: Myobatrachidae) from South Australia. *Copeia* **1989**(2)**:** 488–92.

Roberts J. D., Wardell-Johnson, G. and Barendse, W., 1990. Extended distributions of *Geocrinia vitellina* and *Geocrinia alba* (Anura: Myobatrachidae) from south-western Australia, with comments on the status of *G. lutea*. *Rec. W. Aust. Mus.*. **14**(4)**:** 427–37.

Robertson, J. G. M., 1986. Female choice, male strategies and the role of vocalizations in the Australian frog *Uperoleia rugosa*. *Anim. Behav.* **34:** 773–84.

Robertson, J. G. M., 1986. Male territoriality, fighting and assessment of fighting ability in the Australian frog *Uperoleia rugosa*. *Anim. Behav.* **34:** 763–72.

Shea, G. M. and Johnston, G. R., 1988. A new species of *Notaden* (Anura: Leptodactylidae) from the Kimberley Division of Western Australia. *Trans. Roy. Soc. S. Aust.* **112**(1)**:** 29–37.

Straughan, I. R., 1968. A taxonomic review of the genus *Mixophyes* (Anura, Leptodactylidae). *Proc. Linn. Soc. NSW* **93:** 52–59.

Thompson, M. B., 1981. The systematic status of the genus *Australocrinia* Heyer and Liem (Anura: Leptodactylidae). *Aust. J. Zool.* **29**(1)**:** 93–102.

Tyler, M. J. (ed.), 1983. The Gastric Brooding Frog. Croom Helm: London.

Tyler, M. J., 1984. *There's a frog in my ~~throat~~ stomach*. Collins: Sydney.

Tyler, M. J. and Davies, M., 1985. The gastric brooding frog *Rheobatrachus silus*. Pp. 469–70 *in* The Biology of Australasian Frogs and Reptiles ed by G. Grigg, R. Shine and H. Ehmann. Surrey Beatty & Sons: Chipping Norton with Royal Zoological Society of NSW: Mosman.

Tyler, M. J., Davies, M. and Martin, A. A., 1981. Australian frogs of the leptodactylid genus *Uperoleia* Gray. *Aust. J. Zool.* Suppl. Ser. **79:** 1–64.

Tyler, M. J., Martin, A. A. and Davies, M., 1979. Biology and systematics of a new limnodynastine genus (Anura: Leptodactylidae) from north–western Australia. *Aust. J. Zool.* **27**(1)**:** 135–50.

Watson, G. F. and Martin, A. A., 1973. Life history, larval morphology and relationships of Australian leptodactylid frogs. *Trans. Roy. Soc. S. Aust.* **97**(1)**:** 33–45.

Williamson, I. and Bull, C. M., 1989. Life history variation in a population of the Australian frog *Ranidella signifera*: egg size and early development. *Copeia* **1989**(2)**:** 349–56.

Winter, J. and McDonald, K. R., 1986. Eungella: the land of cloud. *Aust. Nat. Hist.* **22:** 39–43.

Zweifel, R. G., 1972. A review of the frog genus *Lechriodus* (Leptodactylidae) of New Guinea and Australia. *Amer. Mus. Nov.* **(2507):** 1–41.

Family Hylidae

Crossland, M. R. and Richards, S. J., 1993. The tadpole of the Australopapuan frog *Litoria nigrofrenata* (Gunther, 1867) (Anura: Hylidae). *Trans. Roy. Soc. S. Aust.* **117:** 109–10.

Czechura, G. V., Ingram, G. J. and Liem, D. S., 1987. The genus *Nyctimystes* (Anura: Hylidae) in Australia. *Rec. Aust. Mus.* **39**(5)**:** 333–38.

Davies, M., 1989. Developmental biology of the Australopapuan hylid frog *Litoria eucnemis* (Anura: Hylidae). *Trans. Roy. Soc. S. Aust.* **113**(4)**:** 215–20.

Davies, M., Martin, A. A. and Watson, G. F., 1983. Redefinition of the *Litoria latopalmata* species group (Anura: Hylidae). *Trans. Roy. Soc. S. Aust.* **107**(2)**:** 87–108.

Davies, M., McDonald, K. R. and Adams, M., 1986. A new species of green tree frog (Anura: Hylidae) from Queensland, Australia. *Proc. Roy. Soc. Vic.* **98**(2)**:** 63–71.

Davies, M. and Richards, S. J., 1990. Developmental biology of the Australian hylid frog *Nyctimystes dayi* (Günther). *Trans. Roy. Soc. S. Aust.* **114**(4)**:** 207–11.

Hutchinson, M. N. and Maxson, L. R., 1987. Phylogenetic relationships among Australian tree frogs (Anura: Hylidae: Pelodryadinae): an immunological approach. *Aust. J. Zool.* **35**(1)**:** 61–74.

Ingram, G. J. and Hosmer, W., 1982. *Litoria revelata*: a new species of tree–frog from eastern Australia. *Mem. Qld Mus.* **20**(3): 635–37.

Littlejohn, M. J. and Watson, G. F., 1983. The *Litoria ewingii* complex (Anura: Hylidae) in south-eastern Australia. VII. Mating-call structure and genetic compatibility across a narrow hybrid zone between *L. ewingii* and *L. paraewingii*. *Aust. J. Zool.* **31:** 193–204.

McDonald, K. R. and Davies, M., 1990. Morphology and biology of the Australian Tree frog *Litoria pearsoniana* (Copland) (Anura: Hylidae). *Trans. Roy. Soc. S. Aust.* **114**(3): 145–56.

McDonald, K. R. and Storch, D. L., 1993. A new reproductive mode for an Australian hylid frog. *Mem. Qld Mus.* **34:** 200.

Maxson, L. R., Ondrula, D. P. and Tyler, M. J., 1985. An immunological perspective on evolutionary relationships in Australian frogs of the hylid genus *Cyclorana*. *Aust. J. Zool.* **33:** 17–22.

Richards, S. J., 1992. The tadpole of the Australian frog *Litoria nyakalensis* (Anura: Hylidae), and a key to the torrent tadpoles of northern Queensland. *Alytes* **10**(3): 99–103.

Richards, S. J. and Alford, R. A., 1992. Nest construction by an Australian rainforest frog of the *Litoria lesueuri* complex (Anura: Hylidae). *Copeia* **1991**(4): 1120–123.

Richards, S. J., McDonald, K. R. and Ingram, G. J., 1993. Recognition of *Litoria eucnemis* in Australia. *Mem. Qld Mus.* **34:** 94.

Robinson, M. and Cappo, M., 1990 (1989). Comparison of feeding behaviours of the aquatic Australian hylid frogs *Litoria dahlii* (Boulenger, 1896) and *Cyclorana platycephala* (Günther, 1873) and the terrestrial hylid frog *Cyclorana novaehollandiae* (Steindachner, 1867). *Herpetofauna* **19**(1): 8–24.

Tyler, M. J., 1968. Papuan hylid frogs of the genus *Hyla*. *Zool. Verhand.* (Leiden) **96:** 1–203.

Tyler, M. J., 1971. The phylogenetic significance of vocal sac structure in hylid frogs. *Misc. Publ. Univ. Kansas Mus. Nat. Hist.* **19**(4): 319–60.

Tyler, M. J., 1974. The systematic position and geographic distribution of the Australian frog *Chiroleptes alboguttatus* Günther. *Proc. Roy. Soc. Qld* **85**(2): 27–32.

Tyler, M. J. and Davies, M., 1985. A new species of *Litoria* (Anura: Hylidae) from New South Wales, Australia. *Copeia* **1985**(1): 145–49.

Tyler, M. J. and Martin, A. A., 1975. Australian leptodactylid frogs of the *Cyclorana australis* complex. *Trans. Roy. Soc. S. Aust.* **991**(2): 93–99.

Tyler, M. J. and Martin, A. A., 1977. Taxonomic studies of some Australian leptodactylid frogs of the genus *Cyclorana* Steindachner. *Rec. S. Aust. Mus.* **17**(15): 261–76.

Watson, G. F., Littlejohn, M. J., Gartside, D. F. and Loftus-Hills, J. J., 1985. The *Litoria ewingii* complex (Anura: Hylidae) in south-eastern Australia. VIII. Hybridization between *L. ewingii* and *L. verreauxii alpina* in the Mount Baw Baw area, south central Victoria. *Aust. J. Zool.* **33:** 143–52.

White, A. W., Whitford, R. W. and Mahony, M. J., 1994. A new species of *Litoria* (Anura: Hylidae) from Eastern Australia. *Jour. Linn. Soc. NSW* **114**(1): 3–10.

White, A. W., Whitford, R. W. and Watson, G. F., 1980. A redescription of the Jervis Bay tree frog *Litoria jervisiensis* (Anura: Hylidae), with notes on Krefft's frog (*Litoria kreffti*). *Aust. Zool.* **20:** 375–90.

Family Microhylidae

Zweifel, R. G., 1962. A systematic review of the microhylid frogs of Australia. *Amer. Mus. Nov.* **2113:** 1–40.

Zweifel, R. G., 1965. Revisionary notes on Australian microhylid frogs of the genus *Sphenophryne*. *Amer. Mus. Nov.* **2214:** 1–9.

Zweifel, R. G., 1985. Australian frogs of the family Microhylidae. *Bull. Amer. Mus. Nat. Hist.* **182**(3): 265–388.

Family Ranidae

Menzies, J. I., 1987. A taxonomic revision of Papuan *Rana* (Amphibia: Ranidae). *Aust. J. Zool.* **35:** 373–418.

Richards, S. J., 1992. The tadpole of the Australopapuan frog *Rana daemeli*. *Mem. Qld Mus.* **32**(1): 138.

Family Bufonidae

Covacevich, J. and Archer, M., 1975. The distribution of the cane toad, *Bufo marinus*, in Australia and its effect on indigenous vertebrates. *Mem. Qld Mus.* **17:** 305–10.

Easteal, S., 1985. The ecological genetics of introduced populations of the giant toad, *Bufo marinus*. III. Geographical patterns of variation. *Evolution* **39**(5): 1065–075.

Easteal, S., van Beurden, E. K., Floyd, R. B. and Sabath, M. D., 1985. Continuing geographical spread of *Bufo marinus* in Australia: range expansion between 1974 and 1980. *Jour. Herp.* **19**(2): 185–88.

Freeland, W. J., 1986. Populations of cane toad, *Bufo marinus*, in relation to time since colonization. *Aust. Wildl. Res.* **13:** 321–29.

Freeland, W. J. and Kerin, S. H., 1988. Within habitat relationships between invading *Bufo marinus* and Australian species of frogs during the tropical dry season. *Aust. Wildl. Res.* **15:** 293–305.

Freeland, W. J. and Martin, K. C., 1985. The rate of range expansion by *Bufo marinus* in northern Australia, 1980–84. *Aust. Wildl. Res.* **12:** 555–59.

Freeland, W. J., Delvinquier, B. L. J. and Bonnin, B., 1986. Food and parasitism of the cane toad, *Bufo marinus*, in relation to time since colonization. *Aust. Wildl. Res.* **13:** 489–99.

Zug, G. R. and Zug, P. B., 1979. The marine toad, *Bufo marinus*: a natural history resumé of native populations. *Smithsonian Cont. Zool.* **284:** i–iv and 1–58.

Glossary

AMPHIBIA: Grouping of animals (termed a Class) which includes frogs.

AMPHICOELOUS VERTEBRA: A vertebra with a hollow or concave surface at each end.

AMPLEXUS: Sexual embrace of frogs. There are two main positions: axillary in which the male grasps the female at the armpit or around the upper arm, and inguinal in which the female is grasped around the waist.

ANTERIOR: Towards the head end; front.

ANTERO-LATERAL: At the side towards the front.

ANURA: Grouping of animals (termed an Order) which is composed solely of frogs and toads.

ANURAN: A frog or toad.

ARCIFERAL: Shoulder girdle in which the two halves do not fuse in the middle of the chest. (See also firmisternal).

AXILLA: Armpit

BIFID: A structure forked into two portions.

BODY LENGTH: The measurement from the tip of the snout to the cloaca.

BUCCAL CAVITY: The space behind the mouth opening.

CAECILIAN: Legless, worm-like member of the Amphibia included in the Order Gymnophiona.

CANTHUS ROSTRALIS: Area of the head between the eye and the nostril.

CAUDAL: Referring to the tail.

CAUDATA: Grouping of animals (called an Order) which is composed of salamanders and newts.

CHOANA (pl. choanae): Opening of nasal cavity into buccal cavity.

CIRCUM-MARGINAL: Term applied to a groove around the edge of the finger and toe discs of tree-frogs.

CLAVICLE: Collar bone.

CLOACA: Common opening at the rear, for discharge of urine, faeces and reproductive products.

COCCYGEAL: A coccygeal marking or gland is located above the cloaca.

CONSPECIFICS: Members of a single species.

CRETACEOUS: Geological period commencing 135 million years ago.

CRYPTIC SPECIES: Two or more species which are very similar in appearance; usually distinguished most readily by their advertisement calls.

DEVONIAN: Geological period commencing 405 million years ago.

DIGIT: Finger or toe.

DISC: Flattened, expanded and usually circular structure on the tip of a finger or toe.

DISTAL: Away from the body.

DORSAL SURFACE (DORSUM): Upper surface of the back and head.

DORSO-LATERAL: Between the dorsal and lateral surfaces (e.g., dorso-lateral fold).

DORSO-VENTRAL: Between the dorsal and ventral surfaces.

FIRMISTERNAL: Shoulder girdle in which the two halves are fused in the middle of the chest. (See also arciferal).

FLASH MARKING: A patch of vividly contrasting colour on surfaces that are normally hidden (e.g., in the groin or axilla or upon the sides of the thighs).

GLOBOSE: Globe-shaped or round.

GYMNOPHIONA: Grouping of animals (termed an Order) which is composed of legless, worm-like Amphibia.

HORNY JAWS: Hard, black, central jaws of a tadpole. Also called the beak or horny beak.

HERBIVOROUS: Feeding upon plant material.

INGUINAL: In the groin.

INTERCALARY STRUCTURE: Disc or cylindrical rod of cartilage and bone, situated between the last two bones of fingers or toes of some tree frogs (e.g., in *Litoria*).

INTERGRADES: Specimens whose characteristics are intermediate between two morphological races, or two species between which hybridization occurs.

INTERNARIAL DISTANCE: The distance between the nares (nostrils).

JURASSIC: Geological period commencing 180 million years ago.

KERATINIZED: Composed principally of the connective tissue keratin.

LABIAL: Of the lips.

LABIAL PAPILLAE: Small nipple-like projections on the lips.

LABIAL TEETH: Black teeth-like structures between the lips of tadpoles.

LABIAL TOOTH FORMULA: A map-like representation of the number and disposition of tadpole tooth rows, and whether they are continuous or divided in the middle.

LABYRINTHODONTS: An extinct group of Amphibia, possibly ancestral to frogs.

LATERAL: On the side.

LYRATE: Lyre-shaped.

MANDIBLE: Lower jaw.

MAXILLA: Bone contributing most of the upper jaw.

METACARPAL TUBERCLE: A tubercle on the hand.

METATARSAL TUBERCLE: A tubercle on the foot.

MORPHOLOGY: The study of structure.

MIOCENE: Geological period commencing 25 million years ago.

NARIS (pl. NARES): Nostrils — the external openings.

NEUROTOXIC: Having a poisonous effect on nerves and nerve cells.

NOXIOUS: Harmful or injurious to health.

NUPTIAL PAD: Glandular pad usually on thumb of male frogs and either very pale brown or black. Assists the male in securing his grip of the female.

NUPTIAL SPINE: One or more sharp, black spines on the thumb of the male. Function as for nuptial pad.

OMNIVOROUS: Feeding upon plant and animal material.

ORAL: Relating to the mouth.

PAPILLAE: Nipple-like processes.

PAROTOID GLANDS: Raised skin glands located on the top of the head behind the eye.

PECTORAL GIRDLE: Bones and cartilages of the shoulder and chest supporting the arms.

PELVIC GIRDLE: Bones and cartilage of the pelvis supporting the legs.

POSTERIOR: Towards the rear end.

POST-FEMORAL GLANDS: Oval glands on the back of each thigh.

PRE-POLLEX: A small bony structure enlarged on the thumb of some male frogs adjacent to the nuptial pad.

PROCOELOUS: A vertebra concave at one end and convex at the other end.

PROCORACOID: A bone of the pectoral girdle.

RETICULATION: Net-like pattern of markings.

SEXUAL DIMORPHISM: Consistent differences in shape or colour between males and females of the same species.

SHOVEL: A hard, sharp-edged, enlarged tubercle on the undersurface of the foot of many burrowing frogs.

SNOUT-VENT LENGTH: See "Body length".

SPATULATE: Flattened and elongate.

STERNUM: The cartilage or bone comprising the mid-ventral element of the pectoral girdle.

SUPRA-LABIAL: Above the upper lip, e.g., supralabial gland.

TAXON (pl. TAXA): A unit of classification such as "species" or "genus".

TAXONOMY: The study of plant or animal classification.

TIBIAL GLAND: Grossly enlarged, raised poison gland on the dorsal surface of the calf, best known in some *Limnodynastes* species.

TOOTH FORMULA: See "Labial tooth formula".

TRIASSIC: Geological period commencing 230 million years ago.

TUBERCLE: A small rounded or oval lump or protuberance.

TYMPANUM: A taut, circular membrane covering the entire entrance to the ear.

UROSTYLE: A cylindrical bony rod extending posteriorly beyond the vertebrae.

VENTRAL SURFACE: The lower surface of the body and limbs.

VESTIGIAL: Considerably reduced in size.

VOCAL SAC: An inflatable pouch usually beneath the jaws of male frogs, serving as a resonator during call production.

VOMERINE TEETH: Teeth located on the palate.

WEB: Membrane or fleshy skin between the fingers and/or toes.

Index of Scientific Names

Index of Common Names